T0298635

Mechanical Estimating Manual

Sheet Metal, Piping and Plumbing

This page intentionally left blank

Mechanical Estimating Manual

Sheet Metal, Piping and Plumbing

By Wendes Systems, Inc.
Edited by Joseph D'Amelio

River Publishers

Routledge
Taylor & Francis Group

LONDON AND NEW YORK

Published 2020 by River Publishers
River Publishers
Alsbjergvej 10, 9260 Gistrup, Denmark
www.riverpublishers.com

Distributed exclusively by Routledge
4 Park Square, Milton Park, Abingdon, Oxon OX14 4RN
605 Third Avenue, New York, NY 10017, USA

Library of Congress Cataloging-in-Publication Data

Mechanical estimating manual : sheet metal, piping, and plumbing/by Wendes Systems, Inc.; edited by Joseph D'Amelio.

    p. cm.
Includes index.
ISBN 978-8-7702-2373-7 (print) -- ISBN 978-8-7702-2253-2 (electronic)
    1. Buildings--Mechanical equipment--Installation--Estimates--Handbooks, manuals, etc. I. D'Amelio, Joseph . II. Wendes Systems, Inc.

    TH6010.M457 2006
    696--dc22

                          2006041262

Mechanical estimating manual : sheet metal, piping, and plumbing/by Wendes Systems, Inc.; edited by Joseph D'Amelio
First published by Fairmont Press in 2007.

©2007 River Publishers. All rights reserved. No part of this publication may be reproduced, stored in a retrieval systems, or transmitted in any form or by any means, mechanical, photocopying, recording or otherwise, without prior written permission of the publishers.

Routledge is an imprint of the Taylor & Francis Group, an informa business

0-88173-505-1 (The Fairmont Press, Inc.)
978-8-7702-2373-7 ((print))
978-8-7702-2253-2 (online)
978-1-0031-7747-0 (ebook master)

While every effort is made to provide dependable information, the publishe , and editors cannot be held responsible for any errors or omissions.

Quick Reference

This page intentionally left blank

Contents

Section VI—**CONTRACTING FOR PROFIT**

Chapter

Preface

This cost estimating manual, covering labor and material costs for sheet metal, piping and plumbing construction work, will save you time and money, and help get you jobs.

It is a clear, practical, comprehensive mechanical estimating manual, with a tremendous source of valid labor and price data, formulas, charts, graphs, etc. It covers proven methodology, efficient procedures, all types of practical forms, detailed estimating, budget estimating, and has many sample estimates.

It shows you how to produce complete and accurate sheet metal, piping and plumbing estimates quickly and easily. It contains complete man hours, material costs and budget costs. It is clear, concise, comprehensive and clearly organized for easy reference and application to smooth estimating.

It is an indispensable guide and source of data for the various aspects of mechanical estimating, used by contractors, estimators, owners, and anyone involved with estimating mechanical costs on construction projects.

It also is a great aid to supervisors, mechanics, builders, general contractors, engineers, and architects for planning and scheduling work, budget estimating, cost control, cost accounting, checking change orders, etc.

The *Mechanical Estimating Manual* covers:

- The principles of successful estimating and contracting, systematic estimating procedures and sample estimates.

- It covers conceptual and scope budget estimating, as well as fully detailed.

- It covers estimating all types of mechanical equipment; heating and cooling, HVAC units, air distribution, plumbing fixtures, specialties, air pollution and heat recovery equipment.

- It covers estimating all types of sheet metal, pressure pipe and fittings, DWV pipe and fittings, valves, specialties, etc.

- It covers contracting for profit, how to correctly applied overhead and profit markups to bids, how to reduce costs to get jobs, computerized estimating, etc.

Great Need for Valid Labor Figures

There has been a great need for complete and valid man-hour figures on ductwork put together in a practical fashion. Here is a manual that does just that. And combined with complete and valid piping and plumbing labor and cost figures, it thoroughly covers the entire realm of mechanical contracting estimating.

Based on Twenty Years of Cost Records and Time Studies

The sheet metal labor figures contained herein span over twenty years of cost records, labor histories and time studies, not only from the contracting company the author was with during that time period, but also through feedback and verifications from other contractors all over the United States and Canada through seminars, consultation and the Wendes estimating system.

Optimal Estimating Labor by the Piece or the Pound

Estimating galvanized ductwork by the piece was a major undertaking of countless time studies, cost records and labor histories. The time per piece for different sizes and types of ductwork was desperately needed data by the sheet metal industry for decades. It is now fully developed and verified. However, for those that still prefer per pound estimating for sheet metal—both per pound and per piece methods are completely and thoroughly covered herein.

Piping Labor Tables

The piping tables contain valid labor figures which are a consensus of leading successful piping contractors all over the U.S., both large and small—developed and compiled during the past ten years with Wendes Systems piping estimating software owners.

Widespread Application of Manual

If you are a contractor or estimator the manual can improve your estimating accuracy, reduce the time spent estimating, aid you in selective bidding and can help get you more jobs.

If you are the owner of a contracting company or manager of the estimating department, this manual will serve as your set of company standards. The data herein can be adjusted as may be required, to meet your particular productivity rates more precisely. The manual can be used for training and controlling of estimating personnel operations.

If you are a project manager, superintendent, foreman or mechanic, here is as excellent guide for planning, scheduling, and monitoring work.

If you are an engineer, architect, builder or general contractor, this manual will clarify what is all involved in mechanical pricing, and give you quick and accurate budget prices, as well for checking change orders.

Man Hour Labor Tables

The key to estimating labor correctly in the sheet metal and piping industries is to work with man hours rather than dollar costs. Labor is more accurately measured in hours and is more of a constant than dollars. Only after the labor hours are derived for an estimate should they be converted in dollars. Hence, this manual works with man hours productivity rates, which can more directly related to, than dollar figures-by anyone anywhere, at any time, today and next year etc., thus by passing all the different wage rates in the country, the various union and non-union rates and the continual adjustments for wage increases, inflation etc.

Labor Based on Typical Average Mechanic

The labor times recorded in this manual are what the average mechanic, of average competence and training, under average conditions, with typical equipment and tools, with normal motivation and supervision, should be able to do and must in today's market. Where conditions and personnel vary from the typical situation, labor correction factors should be applied. Equipment prices, rentals, budget figures, materials, sub-contractors, etc., are of course, in dollar figures.

Labor Tables Priceless

Since it takes endless years of actual experience, a host of different projects, costly trial and error, a great deal of time, effort, analysis and money to finally reach concrete, valid labor times for the infinite variety of work involved in mechanical work, the labor tables in the manual can, indeed, be priceless to contractors, estimators, engineers, etc.

Manual Directly Applicable to Wendes WinDuct and WinPipe Estimating Systems

Labor tables, price tables, material data, procedures, formulas, etc., all correlate with the Wendes WinDuct and WinPipe estimating systems.

This manual gives an overall perspective and detailed understanding of the estimating systems and contractors can switch from manual estimating using this manual to the computer estimating system quickly and easily.

Section I

How to Prepare Sheet Metal
and
Piping Estimates

This page intentionally left blank

Chapter 1

Successful Estimating Principles

THE CRUX OF SUCCESSFUL CONTRACTING

Successful contracting is built on a foundation of complete and accurate estimates with valid markups for overhead and profit.

That means bids must cover the direct costs of labor, material, equipment and subcontractors, plus the indirect overhead costs of the company as well as providing some profit.

PRIMARY GOALS OF CONTRACTING AND BIDDING

The primary goal of an estimate is to cover all costs as they probably will occur when the job is completed, while still having a reasonable chance of getting the job and of being able to meet acceptable plan, spec and performance criteria.

The primary goal of a markup for overhead costs on bids is to cover the total of the overhead expenditures for the year, and that all the projects contribute their proportionate shares.

The primary goal of a markup for profit is to provide money for capital investments, a return on the corporate investments and as a reward for the hard work and risks. If you can't end up with sufficient earnings on investments commensurate with the burden of contracting, then it might be advisable to put the money into CD's or bonds and take it easy.

The primary goal of contracting is to make money, not lose it. It is not just to keep the company going some way or the other, or keep a lot of people employed.

Preparing complete and accurate bids is the first and most vital step in contracting and survival is predicated on it.

A Valid Estimate Leads to:
* A more positive approach to doing the job well and profitably with everything following more smoothly and positively from the beginning of the contract and on through the completion.

* Satisfied customers and repeat business.

* Fewer call backs.

* Easier to focus on the job, and on meeting plan, spec and performance requirements, rather than on how to skimp to make up for estimate shortcomings.

* Covering all overhead expenses such as rent, office salaries, machinery etc.

* Making a profit, a return on investment, money for capital investments.

* Personal satisfaction.

Poor Estimates Lead to:
* Losing money.

* Risky invalid cost cutting.

* Poorer job quality, loss of repeat business.

* Puts contractor behind the 8 ball before the job is even started.

* More call backs and less money available for them.

* A waste of time and energy business-wise leading to losses or maybe a breakeven situation.

A sale based on a deficient estimate is not a sale—it's a donation or a trade-off at best!

PROBLEMS AND CAUSES OF POOR ESTIMATING

Estimating sheet metal and piping construction projects can be a risky and error prone process and mistakes can and do happen.

The Problems
* Rushing, not allowing or having enough time to do a thorough and accurate job.

* Not all items included that should be.

* Not getting valid, acceptable, competitive equipment quotations.

* Not getting valid, acceptable, competitive sub-contractor quotations.

- Mistakes in math, extensions, summations, summaries, recaps, transferences.

- Wrong quantities of items.

- Missed taking off some ductwork or piping.

- Mistakes in labor calculations.

- Insufficient breakdown of the estimate for the particular project, for adequate estimating accuracy. Too much budgeting, rough pricing for expediency.

- Poor, inadequate overhead or profit markups.

- Not checking estimate thoroughly.

- Inadequate, unclear or incomplete information in plans, specifications, designs, bidding instructions.

- Unpredictable and uncontrollable job site conditions.

- Imperfect or incomplete labor or cost records for estimating purposes.

- Self delusions on prices, labor and markups.

- Incompetence or inexperience of the estimator.

- Negligence or indifference causing poor pricing or label selections.

ESTIMATING COMPETENCE REQUIRED

Hence estimating is a relatively difficult, complicated and judgmental type affair and must be handled with great care by competent, reliable and well trained personnel.

(See the beginning chapters on basics in the sheet metal and piping sections for the requirements of proficient sheet metal and piping estimators.)

Eight Facets of the Estimating Diamond
The Main Categories of an HVAC Estimate

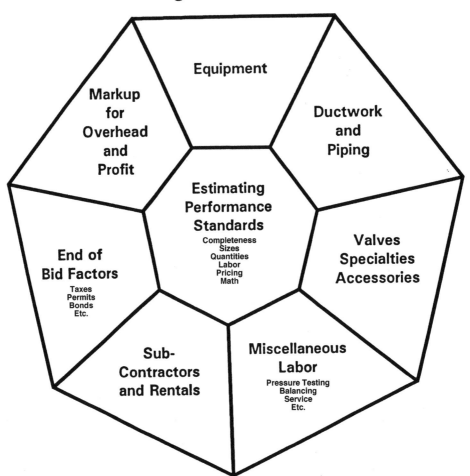

The eight facet categories of estimating must be covered completely and accurately in order to produce valid estimates. Thinking in these terms and keeping them clearly in mind, aids greatly in planning, organizing and remembering everything in an estimate. It's a convenience technique which results in complete and accurate bids.

PERFORMANCE STANDARDS FOR COMPLETE AND ACCURATE ESTIMATES

There are six absolute performance criteria to be met in an estimate.

1. **Completeness**
   - The first and most important goal of a bid is that it is complete. All items must be included that should be and those that shouldn't -are not.
   - All implicit, related and **accessory items**, which are integral parts of a system or normally furnishes as industry practice must be included.
   - The **major categories** of an estimate must be completely covered.
   - **Bastard items**, which vary as to whether you are to furnish or not from job to job, must be thoroughly checked out on every project you bid to determine whether you are to include it or not.
   - Watch out for slippery **end of bid factors** which as permits, bonds, sales tax, items which like bastard items, may or may not be required and must be checked out each and every time.

2. **Based On Valid Sizes or Ranges**
   - Exact detailed sizes is the most accurate basis for labor and material.
   - When size ranges are used for estimating, this must be a valid correlation with the average size in the group. The broader the size group, the greater the risk of inaccuracy.

3. **Quantities Correct**
   Another major goal of an estimate is correct quantities. Do you only have twenty grilles when there should be thirty, or do you only come up with 40,000 pounds or galvanized ductwork when there should be 50,000?

   You only measured 800 linear feet of flexible tubing but there's really 1000 feet. You scaled the sheet metal housing as fifteen feet long and it really needs to be 20 feet long.

4. **Labor Hours Correct**
   Labor is one of the toughest areas to get right and the greatest risk area. It is, however, absolutely necessary to be reasonably correct and to adequately cover the possible variances.

5. **Pricing Correct**
   The pricing on your raw materials, equipment and subs all must be correct. Using $1.25 per pound for stainless steel when it really costs you $2.00 per pound later is a foolish donation to the customer and money lost out of your wallet.

   Make sure you have the right price for the quantity you are working with, higher prices for small quantities and discounted prices for larger amounts.

   Obtain a sufficient number of quotations on equipment and subs so that you know that the price is competitive and not too high or low. Check and compare them for performance and exclusions.

6. **Math Accurate**
   All math operations must be correct, complete, and actually performed. Math errors are subtle and mischievous, and you are not generally aware of them as they occur. A disciplined check is required to ferret them out and to verify their validity.

   Extensions and summations of columns on takeoff, extensions, summary and recap sheets must be correct and complete.

   Adding up labor and material columns on summary sheets is a very vulnerable area. The most drastic errors can occur in the final addition on the recap sheet where can be in 5, 6, or 7 figures.

   The **transfer** of figures, totals etc., from one sheet to another must be done carefully and to perfection. This is an area of frequent error and of cost.

FUNDAMENTAL BIDDING RULES

Allow Enough Time to Properly Prepare Your Estimate

Errors are made due to rushing, pressures, unforeseen interruptions, not making proper takeoffs, not making complete and thorough calculations, not having time to resolve plan and spec errors, omissions, contradictions and vagueness.

Errors are made due to not having time to evaluate equipment and sub prices and to detest and resolve quote problems.

Plan to be done sufficiently ahead of time so there is time for thorough checking, problems, interruptions, for proper digestion and a final realistic perspective view of the job.

Bid Selectively

Only bid on jobs you can do well on and get at a price you can make money on, or at worst, break even on, in rare critical situations.

- You will bid fewer jobs.

- You will get a higher percentage of the jobs you do bid.

- You will cover overhead better and make a profit on jobs.

- You will do better work.

- You will be better able to absorb some of the loses that occur on projects that inevitable occur on estimates or actually on the job.

- You will avoid trading dollars on jobs.

Only bid on projects which you:

- Can get at a price you can live with and that you can be truly competitive on.

- Have experience on or you can perform well on.

- Are adequately set up for in your shop, field, project management, etc.

- Mechanics are sufficiently versed in and perform well on.

- Can effectively work into your construction schedule.

- Have adequate cash flow and operational funds.

- Project within a reasonable geographical area.

- Size of project within your realm of handling.

- Have adequate cost and labor records for accurate estimating.

- Fits into estimating schedule priorities.

Avoid bidding projects with:

- Notoriously low bidders

- Too many bidders

- Where the odds of getting the job are astronomical.

- Against companies who have obvious advantages, experience, personnel, more productive equipment, lower overhead, etc.

Use Uniform Company Estimating Standards

Work with written, substantiated, true company labor productivity rates and material, equipment and sub costs. Have a company estimating manual. Don't let every estimator or yourself pick out their own figures or vacillate with figures from job to job, etc.

Confident and accurate estimating is based on a constant accumulation of historical data market prices. "A price is either current or it is ancient history."

Last year you used $1.25 per pound for stainless steel. You can't assume the price is still the same this year, which may be $1.50

Follow Systematic and Efficient Estimating Procedures

The benefits of a time saving and efficient approach for controlling the preparation of bids are as follows.

- It promotes efficiency and gets bids done faster.
- Duplicating work unnecessarily is avoided.
- It provides time for proper checking of the bid and for solving of problems.
- It provides a frame work for planning and scheduling estimating work realistically and effectively.
- Promotes more complete and accurate bids, more thorough take-offs and more accurate extensions and reliable pricing.
- Through this systematic procedure you will produce more estimates, reduce errors and get more jobs.

Become Thoroughly Familiar with Job Before Bidding

Become Thoroughly Familiar With Job and of the Bidding Requirements, Before Deciding to Bid and Start a Takeoff

Study the plans and specs on a job thoroughly before you decide to bid or start your detailed takeoff.

Analyze the job, the market, your qualifications and ability to compete and make money before making a bidding decision.

Study the plans, specs, bidding instructions and other documents to become familiar with what's involved in the project, what the scope is, what's included and not. Determine what the approximate budget price is, the size of the building and what rough quantities of metal and equipment there are.

Determine if there are alternates or addendum's and what the bidding instructions are.

Become familiar with the areas, floors, systems, equipment, ductwork, conditions, specialties, subs, etc.

Evaluate the competition, architect engineers, generals, agencies, and inspectors involved, cash flow, your work load, the construction schedule, your ability and experience to do the job, your competitive stance, amount of time to bid the job and finally determine intelligently and realistically if you should bid the job or not.

Breakdown Job Sufficiently

Suit the degree of breakdown of the estimate to the degree of estimating accuracy needed on the particular project being bid.

If you just need a rough price for budget purposes, a conceptual budget price with no breakdown at all is adequate. The new 5000 square foot office building will run approximately $40,000 at $8 per square foot. This gives you an accuracy of plus or minus 25%.

If you need more accuracy but not perfection use a semi-detailed budget estimate. Breakdown the system into major component parts and rough budget estimate each part: This increases your accuracy to a plus or minus 15%.

Fans..$0.25 per CFM
Air Handlers.............................$0.75 per CFM
Galvanized Ductwork...............................$3.40 per lb
Louvers ....$20.00 per sq ft

For competitive, firm bidding, which is the bulk of the bidding done, you can't live with being off 10, 15, or 25%. Consequently you must break the job down extensively, not only into its major component parts, but into more specific types, sizes, individual labor operations, and job conditions.

The general rule here is the more you break down the estimate the greater the degree of estimating accuracy.
A corollary to this is the more unfamiliar you are with the item being estimated or the more complicated it is, the greater the need for break down is to minimize inaccuracy

Do Constant Systematic Checking

Human error, distractions, interruptions, lack of sufficient information, unforeseen problems, rushing, all contribute to potential errors as the estimate is being prepared. Systematic, thorough checking is a must.

Select Sub-Contractors Carefully

Subs can frequently make or break a job. You may request only one insulation quote, get a high price and find out after the bids are in that there were lower prices on the street. It is too late then because the job normally would have lost being too high.

Check Equipment Quotations Thoroughly

- Make sure everything is included.

- Know the quantities being quoted on.

- Make sure all components and accessories are included.

- Make sure materials are per design.

- Be aware of exclusions.

- Is the equipment acceptable to the plans and specs.

- Be sure the supplier is quoting a total price for his equipment and not just a unit price.

- Obtain a sufficient number of quotes on equipment and subcontracted work so that you know your price is competitive, neither too high or too low, and complete and accurate.

- Organize and compare the quotations and select the lowest acceptable ones.

Be Acutely Aware of the Right Price for Your Company

The right price for your company and for your level of competition for any job you bid is predetermined before you start your estimate. The right price for your company must truly reflect your operations and the particular job being bid. The price for your operations is a function of

- The productivity of your shop, field, engineering and overhead labor, and your efficiency, machines, tools, facilities, methods, expertise and controls.

- Ability to buy materials, equipment and subs well.

- Your experience and expertise on the particular type building, systems, equipment, ductwork, etc. effects your pricing in relationship to other bidders.

- Know realistically and objectively what overhead percentage you need. If your overhead calls for 30% of the direct costs on a job your bidding based on volume and properly covering indirect costs, and your competitor only needs 20%, you're obviously out of competition assuming your direct costs are about the same.

- The percentage profit desired or required as a return on investment.

- Cash flow available, your financial situation and the cost of money you may have to borrow to handle the project being bid.

- Location of project. The further away you are from the construction site the less control you exercise on it, the higher transportation, delivery, room and board costs are, and the weaker your knowledge and control is of local manpower.

HOW TO ESTIMATE LABOR
ACCURATELY AND REALISTICALLY

There are a number of sources and techniques for deriving sound labor figures for estimating that you can draw from, such as job cost records, time studies, experience, previous estimates and de-tailed break down analysis, etc. as follows:

1. Know How Labor Varies

The factors that make labor vary are:

- Size
- Type
- Material
- Volume
- Duplications

- Number of Labor Operations
- Number of Component Parts
- Productivity of Man Power
- Building Conditions
- Assembled or Broken Down

A single large 50,000 CFM air conditioner broke down into many parts being installed in the penthouse of a forty story building takes a gross amount of labor compared to an assembled AC unit being installed on the first floor of an office building.

2. Cost Records

Labor on previous similar jobs, systems, equipment, ductwork, etc. completed in the past is one of the most valid sources of labor at your disposal.

Your labor record on the previous low rise office building shows you fabricated the low pressure galvanized ductwork at 45 lbs/hr and installed at 24 lbs/hr under normal building conditions. This is vital and usable cost data for your next office building estimate.

3. Time Studies

Time studies, rough spot checks on single items, or group of items is the second most valuable source of labor data. We are not talking about using a stop watch and measuring every motion to the "nth" degree, rather a more general and loose approach.

You ask your mechanic in the shop to keep separate time on a 36"x12" radius elbow he happens to be fabricating and he reports back it took 2-1/2 hours.

You note that a two man crew took 32 man hours to install 100 linear feet of 24 gauge ductwork weighing 730 pounds. This works out to be 26 lbs/hr.

You record the above times on your time study record sheets for future estimating reference. Repeated time studies may be needed of the same items to determine the true average and the range of variation.

4. Previous Estimates

Previous estimates which were prepared in detail and were found to be reasonably in the market range can be yardsticks as to what you subsequent estimating prices should be.

Your last two hospital bids may have run about $12 per square foot of building and the ductwork about $3.70 per pound. These figures can be your guide and comparison for the current hospital you are bidding.

5. Experience

Experience is a vital factor in determining labor not only for the labor times in an estimate but in knowing thoroughly all operations, tools and materials involved.

You recall it took about 24 man hours to install a fan on a previous job. Or you reconstruct in your mind the step-by-step process and approximately how long it took to install a built-up housing on another project.

A consensus of labor times and procedures to perform some work, from a number of people, can turn out to be a very valid source.

6. Detailed Breakdown and Analysis

A detailed breakdown and analysis of an item into all of its component parts and individual operations, for things you're not very familiar with and have no cost records on or which are very complicated, is effective in determining labor.

You break down a kitchen hood into all its parts, tops, sides, front, back, filter rack, and so on, You then calculate the material and labor for each part separately and as well as the assembly labor. In calculating the labor per part you may have to determine what all the sub operations are such as shearing, layout, forming, etc. Set up times may have to be taken into consideration.

7. Correlation and Curves

Make sure your labor times are based on valid correlations. That means that the unit labor used is a true function of whatever the labor is being related to.

The labor to install automatic and fire dampers relates very well to the linear feet of semi-perimeter while the cost of furnishing, the material and fabrication labor corresponds more reliably to square footage.

Round ductwork and flexible tubing correlates to the diameter for installation labor and to the circumference for furnishing costs.

Galvanized ductwork labor corresponds better to the piece of ductwork than to the pound or square foot.

Make quick calculations with curves and cover the entire spectrum of sizes. Cost and labor curves give you a feel as to how costs and labor vary with size etc., graphically portray relationships, help you become familiar with the nature of the cost variations and allow for interpolation and extrapolation.

Curves are relatively simple to work with. A four point plot with the points equally spaced going the entire span provides good accuracy, versatility and a tremendous savings in time. No need to make a time study for every point.

Man
Hours
of
Labor

SIZE:

8.   Use Labor Correction Factors

Coming up with the correct labor on an estimate requires using correction factors to adjust labor up or down for various conditions and requirements.

You start with common denominators for standard conditions and add or subtract percentages for variations such as floors, duct heights, congestion, wide open areas, special spaces, temperatures, existing conditions, local labor and so on.

Ductwork on the 14th floor takes about 20% longer to install than on the first floor to compensate for additional vertical transportation of materials and men.

Thirty-foot-high ductwork requires a 1.3 factor over standard ten foot high ductwork.

Large open areas install faster and standard installation times can be reduced 15%.

9.   Use Valid Labor Averages

Your final objective in estimating labor in a bid is that each component is based on valid labor averages corrected for variable conditions, and that the labor variances up and down will balance themselves out overall so that the total labor is correct in the end.

The range of potential labor variance from the average labor for each item used in the bid must be in a acceptable and reasonable range. The average labor used in a bid must be based on a truly average situation or on a sufficient number of labor studies to make it a valid arithmetical average.

One crew may install a duct run in an average 32 hours, another crew may install the equivalent in 28 hours and a third crew in 36 hours. In the mix you have your valid 32 hour average.

And finally, with all the minute variations in conditions, personnel, equipment and other unpredictable and uncontrollable things in the construction industry, realize that estimating labor is sometimes an approximation or judgment matter rather than an exact science.

DO YOUR HOMEWORK

Keep Cost Records

Keep sufficiently detailed separate cost records on the following items, as a minimum.

*   Costs on equipment, raw materials and subs.

*   Weights on galvanized ductwork, specialties, special metals.

*   Shop labor on galvanized ductwork, specialties and special materials.

*   Field labor on galvanized ductwork, equipment. Indicate productivity rates such as Lbs/Hr on galvanized in both shop and field. Monitor on a monthly basis comparing equipment, material, labor, sub costs, hours, weights and rates with estimate. Analyze final costs and adjust subsequent estimates accordingly.

Keep Up to Date with Your Market and Competitors

Know the approximate dollar volume of construction work in you geographical area of work, and what percentage market penetration you desire and are geared for.

Know how many contractors are competing with you, what their expertise is, the size of their operations, volume of work, bidding and markup strategies, etc.

Be Technically Competent, Knowledgeable and Up to Date

Be knowledgeable about what you are estimating. Know your trade, systems and equipment, how the work is properly done, all the parts needed, what the components and accessories are, the operations involved and type of materials, tools and machinery needed. (See sections on proficient sheet metal and piping estimators.)

USE TIME SAVING ESTIMATING TECHNIQUES

Clarify with Sketches and Diagrams

Draw pictures and diagrams to clarify. Sketch on the plans, on separate sheets of paper or on take off sheets.

Diagram color, write notes, mark whatever is needed on the plans, specs, forms you use.

Plans and specs are all too frequently hazy, incomplete, wordy and need clarification and amplification. Riser sections may be needed. Materials', lining, insulation should be marked on plans. Operations required and component parts not obvious on plans should be indicated. Indicate lengths, quantities, etc. if it aids in your quantity surveys, in your understanding, your memory and organization.

Use Forms

Forms are an indispensable aid and guide to organized, efficient and thorough estimating. They help control the proper sequence of estimating work, continually remind you of what information is needed, lead you logically through calculations and as a result, your bids will be more complete and accurate.

Use Short Cuts

Use short cuts where it is safe to. Reduce tedious takeoff time and excessive extension work, especially if preparing bids manually.

Use a Computer for Speed and Automatic Accuracy

Use a computer for takeoffs, extensions, summaries, recaps, reports, etc. and cut the estimating time on a bids in half or a third-while at the same time greatly increasing the accuracy of the calculations, lookups and generation of valuable information, etc.

Benefits
- Cut **estimating time** in half or a third.

- Perform **lookups** of labor, prices, data with electronic speed and perfect accuracy.

- Perform all the **calculations** for entire jobs automatically and in minutes.

- Make changes in estimates with automatic and instantaneously **recalculations**.

- Print extensive, readable estimating and management **reports** instantly.

- **Focus** on the project and the bidding requirements better.

- Use **formulas, standards, labor** and **price data** which are already built into the computerized estimating system.

Many Problems Disappear with Computer Estimating

Many of the problems that occur in manual estimating automatically disappear with a computerized estimating system, as follows:

- **Rushing** and the time pressure factor is reduced.

- Mistakes in **math.**

- Cumbersome, time consuming **pencil and paper takeoffs.**

- Slow, tedious, error prone manual **lookups** of labor, prices, technical data, etc.

- The messy mass of **manual calculations.**

- The error prone **transfers** of sub totals from sheet to sheet.

- The difficulty of making **changes and recalculations** in estimates.

- Not being able to **concentrate** on the job well enough when bidding manually because of the difficulty of the process.

- The need for extensive estimating reference manuals and **paperwork eliminated.**

Please refer to chapter 22 and computerized estimating for information on the Win-Duct and Win-Pipe estimating systems.

APPLY VALID OVERHEAD AND PROFIT MARKUPS FOR THE JOB AND YOUR COMPANY

Include Valid Overhead Markup

Every job must have a markup that is sufficient to provide it's proportionate share of overhead costs based on the type of job it is, volume of business you are doing and total overhead costs for the year.

Include Profit
- Profit must provide an adequate return or investment, commensurate with other available yields and the risk involved.

- Profit is necessary to buy new machinery, build facilities and other capital investments.

- Profit is necessary as an incentive and reward for hard work, accomplishment and personal satisfaction

Chapter 2

Systematic, Efficient, Accurate Estimating Procedures

BENEFITS

The following is an efficient, systematic, organized, time saving procedure for controlling the preparation of your bids which provides the following benefits:

- It promotes more complete and accurate bids, thorough takeoffs, accurate extensions and reliable pricing.

- It promotes efficiency. You get your bids done faster. You avoid duplicating work unnecessarily. You can get certain things done at the same time following the critical path methodology, which leads to the ultimate shortest amount of time to complete the estimate.

- Bids are more likely to get done on time and thereby allow time for proper checking and solving of problems. Hectic 11th-hour scrambling is avoided.

- It provides a frame work for planning and scheduling estimating work realistically and effectively.

- Through this systematic procedure more estimates will be produced with fewer efforts and you will get the jobs you should and not the ones you shouldn't.

STEPS IN ESTIMATING PROCEDURE

1. **Preliminaries**
This first step of the procedure is a crucial one and it sets the ground work for a proper bid.

The preliminary survey is a systematic, highly organized approach to becoming thoroughly familiar with a job before preparing an estimate and getting into the quagmire of details.

- In the preliminary survey you study the plans, specs and other documents to become familiar with what is involved in the project, what the scope is, what is included and not, what the approximate budget price is, what the size of the building is and what rough quantities of metal and equipment there are.

- You determine if there are alternates or addenda and what the bidding instructions are.

- You become familiar with the areas, floors, systems, equipment, ductwork, conditions, specialties, subs, etc.

- You determine intelligently and realistically if you should bid the job or not by evaluating the competition, architect engineers, general contractors, agencies and inspectors involved, cash flow, your work load, the construction schedule, your ability and experience to do the job, your competitive stance and amount of time to bid the job.

- And lastly you use the preliminary survey as your note sheet and check-off list.

2. **Notify Suppliers**
Immediately after finishing the preliminary survey, notify sub-contractors and equipment suppliers that you will be needing a quotation from them, so they will have adequate time to prepare it, can do so simultaneously as you prepare your bid, and have it ready in time.

Also, make arrangements for any forms needed, pre-qualifications, written proposals, bid bonds, bid deposit checks, etc. so that they are ready at the bid time.

3. **Perform Quantity Takeoffs and Extensions**
Before beginning the takeoff of ductwork and equipment study the plans and specs thoroughly, mark and color the drawings. Highlight different types of duct runs, piping lines and insulated runs in color as required to distinguish one from the other. Locate and mark alternate and addendum areas and conditions that require labor adjustments. Take off major equipment first, then ductwork, piping and small equipment and then specialties.

List everything on the summary sheet, grouping items in the major categories; equipment to start with, then ductwork, piping, specialties, special labor and minor subs.

Price out raw materials, extend shop and field labor and total the labor columns.

4.   Calculate Miscellaneous Labor based on quantity takeoffs and extensions, etc.

5.   Summarize

Enter totals from takeoff extension sheets.

6.   Obtain Supplier Quotations

Call for the quotations that have not come in yet. Make sure they have essential in formation on them such as quantities, types, manufacturers, accessories, exclusions, delivery, do they meet plans and specs, and are materials, sizes, performance correct, etc. Organize and compare the quotations and select the lowest acceptable ones. Plug numbers into summary sheet and total material column.

7.   Obtain Sub Contractor Quotations

Check, compare and select sub-contractor quotations.

8.   Make Thorough Check

Make a thorough check at this point of everything done to this point. Check all takeoffs, extensions, summations, transferences, pricing, labor, etc. Have someone else study project itself and review your estimate. Reread plan, specs, notes, quotes, etc. Have someone else check the math.

9.   Do Recap, Markups, Final Price

Transfer correct totals from summary sheet to the recap sheet. Price out labor and summarize subs. Put in end of bid factors such as sides tax, performance bonds, material and labor increases, contingencies, etc.

Determine the proper markup for overhead and enter. Add everything together and add the desired profit to it. Recheck Recap.

10.   Submit Bid

Submit a proper, qualified bid noting inclusions and exclusions and exceptions to plans and specs.

The above diagram shows a complete, fast and efficient procedure for preparing sheet metal and piping estimates. The diagram shows the correct sequence of operations and the main areas of work. It follows the critical path method showing the sheet metal and piping estimator, HVAC equipment supplier and sub-contractor all preparing their own portions of the estimates at the same time and all coming together for a total bid

price within the bid time frame.

Avoid wasting time and money preparing estimates by locating, identifying and clarifying different duct runs, systems and special requirements before the takeoff is made.

Mark and color drawings before you make your takeoff so that you can easily follow the duct runs and systems for more efficiency, and to not accidentally miss or combine different type items.

Avoid taking off high priced stainless ductwork as much lower priced standard galvanized. Lined ductwork might accidentally be mixed in with the bare galvanized duct takeoff without being identified and then have to be re-taken off to separate it for correct pricing. Alternate areas and correction factor areas may be lumped in with the whole job and then have to be broken out later, doubling the estimating work required.

CHECKING ESTIMATES

Avoiding That Sunken Feeling In Your Stomach

There are many different types of errors that occur in estimating. They are generally made without realizing it at the time, they are made on a rather consistent basis, To avoid losing money and to survive in contracting, you must ferret out the errors and rectify them.

$70,000 is incorrectly estimated for material and labor on a job, instead of $80,000 and $10,000 is lost. A $10,000 markup is put on a job for overhead when it should have been $20,000 and another $10,000 is lost.

Items are left out, counted wrong or added up wrong. It is very difficult to prevent errors 100 percent, but you can methodically and diligently catch them and correct them.

Page 15 shows an example of typical errors made in estimating.

Causes of Errors and Poor Pricing

See Chapter One for the causes of errors and poor pricing in the "Problems of Estimating' section.

Procedure for Avoiding Errors

Avoid crippling loses on bids that are too low or wasting time on those that are too high due to errors by applying the following effective techniques:

1.   The following aspects of an estimate must always be checked at the end of each bid.

   ❑   Are all the **items in**? Is the bid complete?
   ❑   Are the **quantities correct**?

Estimating Procedure Diagram
For Sheet Metal Work And Piping

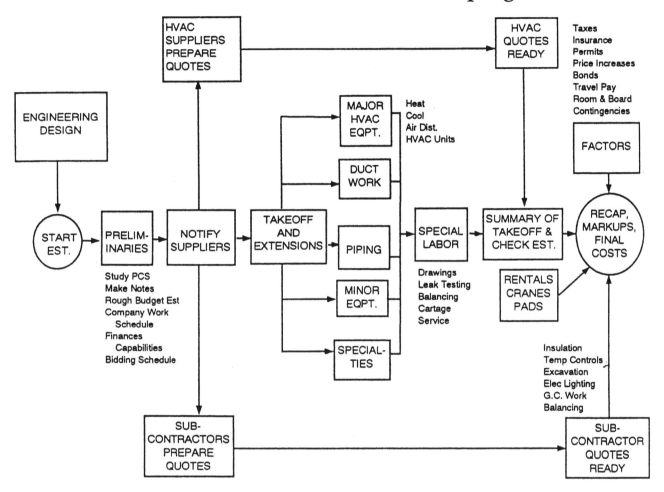

- ❏ **Pricing correct** for equipment, material and subs?
- ❏ **Labor right**? 20, 60, 2000 hours
- ❏ **Math correct**?
- ❏ **Markup** for overhead correct? 20%, 12%, 3 8%

Reread everything and recheck every item one by one.

2. The second most effective step in avoiding estimating errors is to become **thoroughly familiar** with the project before starting the takeoff. Know the systems, equipment, ductwork, conditions, etc. very well, ahead of time.

3. **Don't Rush**! Allow enough time to properly prepare the estimate. If there is absolutely not enough time don't bid, or put in a much higher price than you think it should be. Plan to be done ahead of time and let the estimate digest properly before submittal and commitment.

4. **Constant, Systematic Checking**. Check each item that you take off before you go to the next drawing. Don't carry errors along through a bid. Immediately back check every extension, addition, transference etc. before you move on.

5. **Use a Devil's Advocate**. Have a second qualified person look at the project and check your estimate sheets. Get a certified audit with another set of eyes and viewpoint. Have a committee check it over with you.

6. Do a rough mental **check of all your math** at the end. Then have someone else, a bookkeeper or assistant check in detail on an adding machine for absolute, accurate proof of math validity.

7. **Recheck ductwork weight**. Quickly go over the drawings with a measuring wheel and compare the total linear footage with the totals on your

Identify Different Items, Mark And Color Drawings Before Takeoff

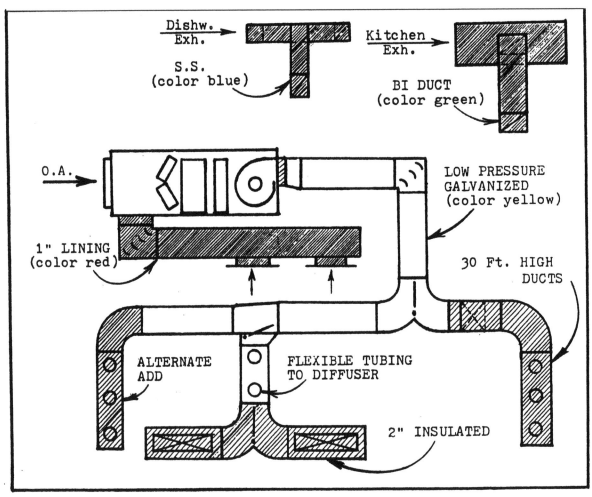

Typical example showing location and identification of duct runs.

takeoff sheets. Compare the average weight per square foot of building with budget figures to see if they are reasonably close.

8. Check totals, unit prices and specific figures against **budget figures, past jobs, previous estimates** and cost records. How does the cost per pound, per ton or per square foot of building compare?

9. **Check your recap**, a horrendous place to make an error. Are the numbers transferred correctly, is the math right, are wage rates correct and are taxes, permits and bonds included. Are contingencies, risks and wage and material price increases covered?

10. Objectively **recheck your overhead markup**. What good is it if you get all your labor and material costs correct and blow it on the markup?

What's your yearly overhead and what must this job contribute to it?
Are you deluding yourself because you want this job badly?
What's the material/labor ratio?
What risks are involved?

11. Eleven can be your lucky number if you properly checked your bid... however, **if you are in doubt or the risks are very high** on a particular project, then consider not bidding. Why roll the dice and come up with a two or twelve and crap out? The risks outweigh the gain tenfold.

Check Quotations Thoroughly

• Make sure everything is included. Know exactly what is being quoted on and to what extent it is be-

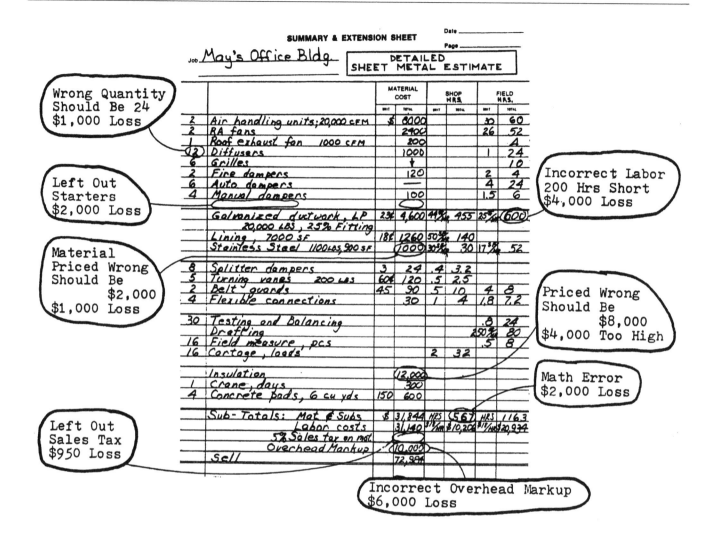

ing covered. A supplier may be quoting all the steel fans but none of the PVC ones, and he may not state this.

- Know the **quantities** being quoted on. An air handling unit company may only be quoting seven units instead of the nine really required and not indicate so. This could cost an extra $4,000.

- Make sure all **components and accessories** are included. Don't find out after the bid has been accepted that the fan quote did not include $3,000 worth of inlet dampers that you are responsible for.

- If the fan wheels must be aluminum, be careful not to quote based on steel wheels. Make sure **materials** are per design.

- Note if the equipment being quoted on is to be **shipped assembled** or **knocked down**. The manufacturer may have a personal money saving plan to

send the complex air handling units broken down into components for you to assemble on the job site causing many extra hours for you not covered in the bid.

- Be aware of **exclusions**.

- Is the equipment acceptable to the **plans and specs**?

- Be sure the supplier is quoting a **total price** for his equipment and not just a **unit price**.

- Obtain a sufficient **number of quotes** on equipment and subcontracted work so that you know your price is competitive, neither too low or too high, and complete and accurate.

- Organize and compare the quotation and select the **lowest acceptable** ones.

Scope of Complete Sheet Metal Estimate
Check-off List

AIR HAND. EQUIP.
HVAC Units
- ☐ Air Handling Units
- ☐ Coils, Filters, Dampers
- ☐ Roof Tops,Economizers
- ☐ Make Up Air
- ☐ Heat Recovery

Fans
- ☐ Centrifugal
- ☐ Vent Sets
- ☐ Tubular
- ☐ Industrial
- ☐ Propeller
- ☐ Isolators
- ☐ Inlet Vanes
- ☐ Drives, Motors
- ☐ Access Doors

Roof Equipment
- ☐ Roof Exhaust Fans
- ☐ Gravity Vents
- ☐ Louvered Vent Houses
- ☐ Roof Curbs

Filters
- ☐ FG Throw Away
- ☐ Washable Mesh
- ☐ Rollomatic
- ☐ Electro Static
- ☐ Bag
- ☐ Absolute
- ☐ Charcoal

Air Diffusion Equipment
- ☐ Grilles
- ☐ Diffusers
- ☐ Linears
- ☐ Troffers
- ☐ Extractors
- ☐ Slot Diffusers

Terminal Equipment
- ☐ VAV Boxes
- ☐ Dual Duct
- ☐ Reheat
- ☐ Induction

Industrial Exhaust
- ☐ Auto, Welding Exhausts
- ☐ Dust Collectors
- ☐ Scrubbers
- ☐ Paint Spray Booths

Dampers
- ☐ Manual
- ☐ Motorized
- ☐ Fire, Smoke

Electrical
- ☐ Starters
- ☐ Heating Coils

Factory Fab. Housings
Sound Attenuators
Thermometers

DUCTWORK
Galvanized
- ☐ Low,Medium,High Prs
- ☐ Spiral

Fiberglass
Round HVAC
- ☐ Flex Tubing
- ☐ Resid. Furnace Pipe
- ☐ Flues
- ☐ Collars

Industrial Exhaust
- ☐ Black Iron
- ☐ Stainless Steel
- ☐ Aluminum
- ☐ Blow Pipe
- ☐ FRP
- ☐ PVC
- ☐ PVC Coated Galvanized
- ☐ Transite
- ☐ Metal Flex. Tubing

Lining
- ☐ F.G. Flexible
- ☐ Hardboard
- ☐ Cement and Pins

SHT MTL SPECIALTIES
Duct Accessories
- ☐ Turning Vanes
- ☐ Splitter Dampers
- ☐ Bracing Angles
- ☐ Cleat, Hangers
- ☐ Trapeze Angles
- ☐ Spanning Angles
- ☐ Crossbreaking
- ☐ Seal Ducts
- ☐ Paint Ducts

Specialties
- ☐ Flexible Connections
- ☐ Belt Guards
- ☐ Hsg. Access Doors
- ☐ Exh Hoods-Kit, Lab,Shop
- ☐ Stands, Platforms
- ☐ Drain Pans
- ☐ Blankoffs, Safeoffs

- ☐ Roof Hoods
- ☐ Screens, Grating
- ☐ Expanded Metal
- ☐ Perforated Plates
- ☐ Water Eliminators
- ☐ Lead, Cork, Foam Glass

Sheet Metal Housings
- ☐ Panels
- ☐ Angles

SPECIAL LABOR
- ☐ Cartage
- ☐ Field Measuring
- ☐ Drafting
- ☐ Testing and Balancing
- ☐ Leak Testing
- ☐ Temporary Heat
- ☐ Set Up and Clean Up
- ☐ Chases and Sleeves
- ☐ Existing Buildings
- ☐ Removal
- ☐ Cut Openings, Patch
- ☐ Protection

SUB CONTRS, RENTALS
- ☐ Cranes, Hoists
- ☐ Concrete Pads
- ☐ Scaffolding
- ☐ Testing and Balancing
- ☐ Insulation
- ☐ Temperature Control
- ☐ Refrigeration, AC
- ☐ Heating
- ☐ Electrical
- ☐ Cut Openings, Patch
- ☐ Excavate, Backfill

END OF BID FACTORS
- ☐ Sales Tax
- ☐ Permits
- ☐ Travel Pay
- ☐ Room and Board
- ☐ Wage Increases
- ☐ Material Increases
- ☐ Premium Time
- ☐ Alternates
- ☐ Contingencies
- ☐ Clean Up Charges

MARKUPS
- ☐ Overhead
- ☐ Profit

Heating Equipment
Check-off List

Boilers
- ❏ Cast Iron
- ❏ Steel Shell
- ❏ Scotch Marine
- ❏ Gas Fired
- ❏ 09 Fired
- ❏ Electric
- ❏ Combination Gas & Oil
- ❏ Coal, Wood etc.
- ❏ Hot Water
- ❏ Steam

Burners
- ❏ Gun Type
- ❏ Impingement Jet
- ❏ Flame Retention Oil

Draft Controls
- ❏ Barometric
- ❏ Vent Dampers
- ❏ Induced Draft Fan

Cons
- ❏ Hot Water
- ❏ Steam
- ❏ Electric
- ❏ In Air Handling Unit
- ❏ In Duct
- ❏ In Sheet Metal Housing

HVAC Central Units
- ❏ HV AHU
- ❏ HVAC AHU
- ❏ Roof Top
- ❏ Make Up Air Units
- ❏ Furnaces

Heating Terminal Units
- ❏ Fan Coil Units, Cabinets
- ❏ Induction Units
- ❏ Unit Heaters
- ❏ Duct Heaters
- ❏ Baseboard, Fan Tube
- ❏ Baseboard, Radiation
- ❏ Radiators
- ❏ Infrared Units
- ❏ Air Curtain Heaters
- ❏ VAV Boxes
- ❏ Constant Air Volume Boxes

Specialties
- ❏ Steam traps
- ❏ Steam Condensate Meter
- ❏ Separators (Entrainment Eliminator)
- ❏ Vacuum Breakers
- ❏ Expansion Tank
- ❏ Automatic Air Vent
- ❏ Aerators
- ❏ Water Level Controls
- ❏ Water Treatment
- ❏ Other Valves (see piping & valves)

Pumps
- ❏ Centrifugal
- ❏ Condensate
- ❏ Feed Water
- ❏ Smaller Impeller
- ❏ Install Smaller Pump
- ❏ Install Smaller Motor

Flues, Breechings
- ❏ Flue
- ❏ Breeching
- ❏ Factory Fabricated Stack

Cooling Equipment Check-off List

Chillers
- [ ] Centrifugal
- [ ] Reciprocating
- [ ] Water Cooled
- [ ] Air Cooled
- [ ] Package Type
- [ ] Remote Condenser
- [ ] Heat Recovery Reciprocal
- [ ] Double Bundle Condenser

Condensers
- [ ] Air Cooled
- [ ] Water Cooled

Cooling Towers
- [ ] Packaged Gravity
- [ ] Packaged Forced Flow Galvanized
- [ ] Crossflow Induced Field Assembled
- [ ] Forced Draft Field Assembled

Central HVAC Units
- [ ] HVAC Air Handling Units
- [ ] Roof Top Units
- [ ] Self Contained AC Units
- [ ] Air Cooled
- [ ] Water Cooled

Cooling Terminal Units
- [ ] Terminal Under Window Units, Chilled Water
- [ ] Induction Units
- [ ] VAV Boxes (see VAV conversions)
- [ ] Constant Air Volume Boxes (see air handling equipment)

Specialties
- [ ] Receivers
- [ ] Valves (see piping and valves)
- [ ] Thermal sion Valves

Coils
- [ ] DX Coils
- [ ] Chilled Water Coils
- [ ] Reheat Coils
- [ ] Heating Coils
- [ ] "A" Coils
- [ ] Electric Coils

End of Bid Factors to Consider Check-off List

- [ ] **Sales Tax**
  Required on material and equipment on private work which you are selling directly to a user. Not required on government projects for non profit organizations or jobs for resale.

- [ ] **Performance Bonds**
  Can run to 2% of bid price.

- [ ] **Permits**
  Check local codes on heating, refrigeration, ventilation, air pollution, requirements.

- [ ] **Wage Increase**
  Determine schedule, union wage increases, calculate calendar span of job being bid and add for wage increases if they occur.

- [ ] **Material Price Increases**
  Determine when contracts will be awarded, when equipment purchases will be made, what price trend is on raw materials, galvanized, fiberglass, etc. and when you will need them. Add percent increase if anticipated.

- [ ] **Travel Pay**
  Include travel pay per mile for traveling to job site if outside boundaries of local union jurisdiction, as required by contract.

- [ ] **Room and Board**
  If you have to send your people to the job to stay include required room and board costs.

- [ ] **Clean Up Charges by General Contractor**
  Check specs on house cleaning charges, if pro rated, who is responsible. Determine if G.C. is one who slaps subs with clean up charges.

- [ ] **Alternates**
  Make sure additive and deductive alternates are included, especially on government projects, where your bid may be rejected due to bidding irregularity.

- [ ] **Unit Prices**
  Base unit prices on small quantities, on no discount

situations, on low productivity rates, on heavy over-head involvement.

❒ **Special Contingencies**
Consider special contingencies or risks that may occur on the particular job being bid, such as labor availability or quality, adverse job site conditions, tight schedule involving extra costs due to crashing job.

Remodeling Work Factors
Check-off List

Working on HVAC systems in existing buildings takes a great deal more time than it does on new buildings. It's riskier, involves more complications and requires that many additional factors be considered when estimating or scheduling.

❒	OPENINGS TO CUT	Do openings need to be cut? Do they have to be patched?
❒	CEILINGS TO REMOVE	Do ceilings have to be removed? Who replaces ceilings? Do entire ceilings being removed or just cut where needed?
❒	REMOVAL OR EXISTING HVAC EQUIPMENT, ETC.	Fans, pumps, HVAC units, ductwork, piping, etc?
❒	PROTECTION	Does owner's furniture or equipment have to be protected?
❒	CEILING SPACE AVAILABLE	What ceiling space is available? Where are beams located?
❒	EQUIPMENT ROOM SPACE AVAILABLE	Is there space for the new HVAC equipment, ductwork, piping, etc?
❒	OBSTRUCTIONS TO AVOID	Piping and ducts Walls and partitions Lights and conduit Beams, joists and columns
❒	OCCUPIED AREAS	Will areas be occupied during installation?
❒	SHUTTING DOWN OF SYSTEMS	Must systems be in operation or can they be shut down during installation?
❒	SEQUENCE OF WORK	In what order must work be done?
❒	WORKING TIMES	Week days, evenings, weekends. Over time.
❒	MATERIAL HANDLING AND HOISTING	Elevators available Dock available Door, window, wall openings available and large enough for moving equipment through.
❒	CLEAN UP AND SCRAPPING	Material handling labor, dumpster rentals, scavenger services.

15 Bastards with No Regular Homes Check-off List

Responsibility for furnishing these 15 bastard items keep switching from one part of the specs or contract documents to another and back again, and can drive sheet metal and piping contractors and others right up the wall. These are items that may be part of your contract on one job and not on the next.

- Responsibility for furnishing an item can be found in any of six different parts of the specifications on six different projects.

- The items required in your specifications may show up on any of the five different sets of drawings: architectural, lab equipment, piping, kitchen equipment, or ventilation.

- Contract documents may not say explicitly "furnish" or "don't furnish."

The 15 bastards that evade regular assignment in sheet metal and piping bids are as shown in the following check-off list:

❑	Starters, Final wiring
❑	Louvers
❑	Coils in air handling units
❑	Electric duct coils
❑	Rooftop units
❑	Air system testing and balancing
❑	Breechings, stacks and flues
❑	Insulation
❑	Finish painting
❑	Concrete pads
❑	Roof curbs, Duct chases
❑	Wall and ceiling access doors
❑	Exhaust hoods

The Problems That Result

The Industry's approach of randomly assigning responsibility for furnishing and installing these components creates complications and confusion for bidders; it leads to omissions, duplications, overlapping, and errors; it creates contingency allocations in bids.

Haphazard assignment of responsibility and location on drawings wastes time, money and energy. It causes delays, battles, district, lawsuits, exasperating wrangles, dissatisfied customers, and even bankruptcy.

With unclear, indefinite plans and specs, the owner may pay twice for items, because two contractors want to cover themselves in this risk situation and both put money in their bids for the item.

Or no contractor includes money in his bid for an item-let's say access doors in walls and ceilings and when the moment comes for these to be installed near the end of the job and the owner ends up paying more for the item than he would have at bid time as an extra.

Or one contractor furnishes the item-free because he is forced to, based on "It's implicit in the contract documents," etc.

A contractor can lose $5,000 here on starters, $4,000 there on finish painting and $3,000 somewhere else on final wiring without blinking.

An owner may pay $5,000 twice for the same louvers or $2,000 to two different contractors for installing the same electric heating coils.

Contributing Factors

Obviously, the number one reason why these items have no regular homes is that the construction industry is neither uniform nor consistent in assigning responsibility for furnishing and installing specific items.

The second culprit is vagueness or hedging, not being definite or clear as to who should furnish an item, who should install it, and who should pay for installation labor.

General contractors cause some of this chaos when they are the primes, chopping up the specs differently from how they were written. A general may take insulation out of the heating and ventilating sections and sub it himself He may want to furnish the louvers himself even though they are in the ventilation specs.

Confusion, ignorance, and indifference regarding union jurisdiction over who does what and contractor association agreements on who furnishes what contribute to the difficulty in properly grouping and assigning all items involved.

Often a lack of knowledge by the preparer of the contract documents as to how things are installed most economically, what the various contractors actually do, hampers uniformity and consistency in spec writing and drawing.

Time pressures, convenience, self interest, the changing industry, new or untrained personnel, and costs may influence A-Es in their preparation of construction documents.

Lack of coordination among the architect, structural engineer, mechanical engineer, electrical engineer, and spec writers adds to the dilemma.

Operations Included in
Labor Hours

All man hour figures in this manual for everything, ductwork, accessories, equipment, etc. include all direct and indirect labor operations as follows:

AD Shop Labor for Sheet Metal Includes:

1. Unloading and storing raw materials
2. Listing, shop tickets, shear sizes
3. Getting materials from stock
4. Blanking
5. Layout and cutting
6. Forming
7. Seaming
8. Assembling
9. Cleats
10. Hangers
11. Reinforcing angles
12. Loading assembled items on truck
13. Idle and non-production time
14. Supervision time

All Field Labor Includes:
1. Unloading and distribution
2. Unpacking
3. Set up scaffolding, tools, ladder
4. Layout of hangers, etc.
5. Actual hanging of ductwork and piping, or setting up of equipment
6. Tear down of scaffolding, etc. and clean up
7. Idle and non-production time
8. Supervision time

The Following Items are Not Included in the Regular Labor Tables
Field Sketching
Shop Drawings
Truck Driver Labor

This page intentionally left blank

Chapter 3

Sample HVAC Estimate and Forms

OVERVIEW OF SAMPLE JOB

Sample HVAC Estimate The following sample estimate is of a small 6,800 sq ft office building with a low pressure, single zone galvanized duct system, an air handling unit and a split DX system. It includes most of the typical air distribution components in it involved in most projects such as rectangular and round galvanized ductwork, the air handling unit, the roof exhaust fan, dampers, louvers, grilles, registers and diffusers, duct wrap, acoustic liner, duct accessories, etc.

Treat as if you are the prime HVAC contractor, doing the air distribution portion while subbing out the refrigeration and heating pans.

PURPOSE OF FORMS

Forms are an indispensable aid and guide to organized, efficient and thorough estimating. They help control the proper sequence of estimating work, continually remind you of what information is needed, lead you logically through calculations and as a result your bids will be more complete and correct.

Job Description and Budget Costs Form
1. Budget estimate prices to determine if it should be bid or not, and as a check price against the detailed estimate after the bid is complete.

2. Approximate heating, cooling and CFM loads and rough out ductwork weight for check on detailed ductwork takeoff.

3. Record the key characteristics of the type of system involved.

Per Piece Duct Takeoff Sheet
Estimating ductwork labor by the piece is the most accurate and clearest method available for contractors. The takeoff involves listing the duct size, type, quantities on fittings and lengths on straight duct.

The extension of material involves totaling footages per line, entering the weight per running foot and multiplying for the total material weight on each line. The extension of labor involves totaling the quantity of pieces, looking up and entering labor hours per piece for the shop and field and multiplying out for the totals per fine. After the fines are extended the columns are totaled.

Quantity Takeoff Sheet and Extension Sheet
The quantity takeoff sheet is a general form for taking off and listing types, sizes, quantities, etc. of the various items required in a bid other than ductwork or piping, for extending the material amounts, labor, costs, etc. and summations.

Estimate Summary and Extension Sheet
The summary sheet is used as a fine item summary of all the major grouping of different items included in the estimate, from ductwork and other takeoff sheets, etc. It should be divided into the major divisions of a bid, quoted equipment, ductwork, piping, specialties and accessories, miscellaneous labor, etc.

The total amounts of material quantities, labor, etc. are transferred from duct and piping takeoff sheets, quantity takeoff sheets etc. to this summary sheet.

Bid Recap Sheet
1. Recap the job totals of direct costs on labor and prices on raw materials, equipment and sub contractors, and to total them.

2. Put markups on each group and total the overhead markup.

3. Put a profit markup on the labor, raw material, equipment and sub-contractor groups.

4. Total everything for a bottom line bidding price.

Calculating Labor Costs Per Hour
This form insures that all the components of the wage rate which include, base wage rate, normal union

fringe benefits, federal and state payroll taxes, insurance's and dues, are covered in the rate used in a bid.

Telephone Quotations Form

The telephone quotation form is for recording quotations which come over the phone, in an organized, complete and readable fashion. It includes a check-off list on the bottom of critical aspects of a quote such as, if they meet plans and specification requirements, addendum's,

taxes, freight, lead times, etc. A box is provided for exceptions on what is not included.

Bidding Record Form

The purpose of the bidding record form is to have a written record of who the phone bid were given to, what the amount was, what the inclusions and exclusions were and what the plans, specifications and addenda of the bids were based on.

Specifications on Sample Job

IBM Sales Office

A. Related Documents

The general provisions of the contract, the general conditions and supplementary conditions of these specifications plus the A/A document A201-1976 "general conditions" apply to the work in this specification.

B. Scope of Work

Scope of work to include, but not be limited to the following:

Equipment
1. Air handling unit
2. Roof exhaust fans
3. Grilles and registers
4. Ceiling diffusers, lay in
5. Fire dampers with sleeves
6. Control dampers
7. Louvers

Ductwork
8. Galvanized ductwork, low pressure
9. Spiral
10. Flexible tubing
11. Flues

Sheet Metal Specialties
12. Turning Vanes
13. Splitter and manual volume dampers
14. Access Doors
15. Flexible Connections

Insulation
16. Liner, RA duct, 1," 2 lb density
17. Duct wrap, OA Duct, 1- 1/2," 3/4 lb density

Miscellaneous
18. Prepare blown up shop drawings
19. Test & balance air distribution system
20. Refrigeration piping, valves, condenser, insulation
21. Temperature Controls
22. Insulation

C. Work Not Included
1. Painting
2. Power wiring to mechanical equip.
3. Structural steel openings
4. Condensate piping
5. Gas Piping
6. Concrete Pads
7. Starter

IBM Offices, Drawing M-1
Low Pressure, Galvanized, Cleat Connections, SZ
Split DX System

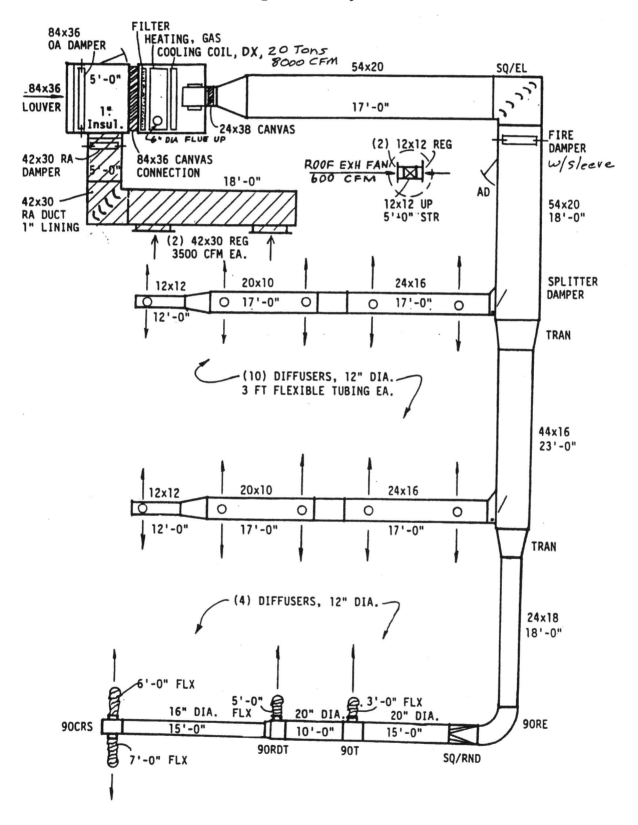

Job Description and Budget Costs

Job ___IBM Offices (Low Rise)_____ Date _____

Location _____ Distance ____10____ Miles

Total Project Costs $ _____ Volume of Building _____ Cu Ft

Total Area __6800__ Sq Ft,   Area₁ _____ Sq Ft   Area₂ _____ Sq Ft

BUDGET COSTS

	COST/SQ FT BLDG	TOTAL	COST/TON	TOTAL
Total HVAC 6800 sq ft	$ 7.00	$ 47,600	$ 2,3.80	$ 47,600
Sheet Metal 3400 lbs	$ 3.50/lb	$ 11,900	$	$
Piping	$	$	$	$
Equipment	$	$	$	$
Insulation	$	$	$	$
Temperature Control	$	$	$	$
Electric	$	$	$	$

DESIGN LOADS

	Area₁		Area₂	
	Factor	Total	Factor	Total
Cooling	340 Sq Ft/Ton	20 Tons	Sq Ft/Ton	Tons
Cooling	35 BTU/Sq Ft	238,000 BTU	BTU/Sq Ft	BTU
Heating	35 BTU/Sq Ft	238,000 BTU	BTU/Sq Ft	BTU
Supply Air	1.1 CFM/Sq Ft	7,480 CFM	CFM/Sq Ft	CFM
Duct Weight	.5 LBS/Sq Ft	3,400 LBS	LBS/Sq Ft	LBS

SYSTEM DESCRIPTION

Heating Gas
Cooling Split DX
☒ SZ   ☐ MZ   ☒ Constant Volume   ☐ VAV
Duct Pressure   ☒ LP   ☐ MP   ☐ HP
Return Air Method   ☒ Duct   ☐ Ceil. Plenum
Type Outlets   Diffusers
Type Perimeter Heat   ☒ Air   ☐ Baseboard
Temperature Control   Electric

KEY PLAN

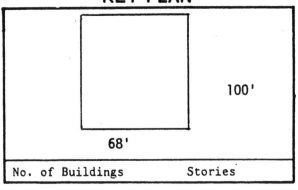

100'

68'

No. of Buildings	Stories

CONSTRUCTION: Glass _____ Gross Area _____ Sq Ft; U_____

Exterior Walls _____ Gross Area _____ Sq Ft; U_____

Roof _____ Gross Area _____ Sq Ft; U _____

Remarks _____

_____

Per Piece Duct Takeoff Sheet
Galvanized

Job __IBM Office__ _____ Drawing _____ System __S-1__ ☐ Lining __1"__

Type Duct: ☒ Galv, ☒ LP ☐ HP, ☐ Other __Rectangular__ _____ ☐ Insulation _____

DUCT SIZE	TYPE DUCT	EQUIVALENT LINEAR FEET PER PIECE		TOT LF	WEIGHT		QTY	SHOP LABOR		FIELD LABOR	
					LBS LF	Total		Hrs /Pc	Total	Hrs /Pc	Total
		STRAIGHT									
54 x 20	STR	17-18		35	21.0	735	8	.9	7.2	2.9	23.2
44 x 16	"	23		23	17.0	391	5	.8	4.0	2.3	11.5
24 x 18	"	18		18	9.8	176	4	.4	1.6	1.1	4.4
24 x 16	"	17-17		34	9.3	316	8	.4	3.2	1.1	8.8
20 x 10	"	17-17		34	7.0	238	8	.3	2.4	.9	.7.2
12 x 12	"	12-12		24	4.4	106	6	.2	1.2	.5	3.0
84 x 36	"	5		5	40.0	200	1	1.3	1.3	4.4	4.4
44 x 32 L	"	5-18	(13 sq/ft)	23	21.5	495	4	.9	3.6	2.6	10.4
12 x 12	"	5		5	4.4	24	1	.2	.2	.5	.5
		STR TOTALS		201		2,681	lbs		24.7		73.4
		FITTINGS									
54 x 20	TRN	3-3	Supply	6	21.0	126	2	2.7	5.4	2.7	5.4
"	SQN	5		5	21.0	105	1	3.3	3.3	3.8	3.8
44 x 16	TRN	3		3	17.0	51	1	2.2	2.2	2.2	2.2
24 x 18	90L	6		6	9.8	59	1	1.5	1.5	1.3	1.3
"	SQRD	3		3	9.8	29	1	4.6	4.6	1.2	1.2
24 x 16	TRN	3-3		6	9.3	56	2	1.0	2.0	1.2	2.4
"	TAP	1-1		2	9.3	19	2	.6	1.2	1.2	2.4
20 x 10	TRN	3-3		6	7.0	42	2	.8	1.6	1.0	2.0
44 x 32 L	TAP	1-1-1	(13 sq/ft)	3	21.5	64	3	1.4	4.2	2.5	7.5
44 x 32 L	SQL	4	(13 sq/ft)	4	20.4	82	1	2.9	2.9	3.8	3.8
12 x 12	TAP	1-1		2	4.4	9	1	.4	.4	.5	.5
		FITTING TOTALS		56		670			32.9		32.5
		TOTAL STR & FITT.		257		3,343	lbs.		58 HR		106 HR
		DUCT ACCESSORIES									
56 x 20	TV	1	QTY			40	1		1.1		---
42 x 30	TV	1				60	1		1.2		---
24 x 18	SPD	1-1					1		2.0		
84 x 36	CANV	1									
24 x 38	"	1									
12""O	FLXT	10 @ 3-6-7-5-3		51		---	14		---	.7	9.8
LINING		299 + 39 + 52 = 390 x		1.15	=	447 sq ft					

Quantity Takeoff Sheet

Job **IBM Offices**

DESCRIPTION
Small Equipment

Drwg Flr or Sys	Size	Description	Quantity	Total	LABOR HOURS	MATERIAL COSTS QUOTE		
M-1	84×36	Louver	1		5.0			
	56×20	Fire Damper	1		2.6			
	84×36	OA Damper	1		4.3			
	42×30	RA Damper	1		2.6			
	12" ⌀	Diffusers	14	1	14.00			
	42×30	Registers	2	1	2.0			
	12×12	Registers	2	1	2.0			

Estimate Summary and Extension Sheet

Job IBM Offices
1 Story, 6,800 Soft

IBM OFFICES

	EQUIPMENT	MATERIAL COST		SHOP		FIELD	
		UNIT	TOTAL	UNIT	TOTAL	UNIT	TOTAL
1	AHU, 20 TONS, 8000 CFM, DX, GAS, 2 STAGE	$	9,953		---		18.0
1	ROOF EXH. FAN, 600 CFM		600		---		2.0
1	FIRE DAMPER, 56 x 20		110		---		2.6
1	LOUVER, ALUM. 84 x 36		465		---		5.0
1	OA DAMPER, 84 x 36		----		---		4.3
1	RA DAMPER, 42 x 30		----		---		2.6
14	DIFFUSERS, FIXED, 12"Ø, LAY IN, SQ.		340		---	1	14.0
4	REGISTERS (2) 42 x 30, (2) 12 x 12		400		---	1	4.0
	SUB TOTAL EQPT.	$	11,868				
	DUCTWORK						
	LP GALVANIZED 3330 LBS	$ .38	1,265		58.0		106.0
	LINING 447 SQ FT	.44	197	50	9.0		---
	SPIRAL 285 LBS	1.20	342		---		11.0
4	FLUES		100			.8	32.0
14 PCS	FLEXIBLE TUBING 12"Ø, 51 FT	2/FT	102		---	.7	9.8
10	COLLARS, CLAMPS, 12"Ø, $3 + $1	4	40		---	.5	5.0
	SHEET METAL SPECIALTIES						
2	TV, 56 x 20, 42 x 30 100 LBS	$ .75/LB	75		2.3		---
2	SPLITTER DAMPERS, 24 x 18, 14 LBS ea	6 ea	12	.5	1.0		---
2	REACH THRU ACC DOORS	15 ea	30		---	.5	1.0
2	FLEXIBLE CONN. 84 x 36, 24 x 38, 30 ft	.80	24	1.0	1.9	2.6 1.6	4.2
	SUB TOTAL MATERIAL	$	2,187				
	SPECIAL LABOR						
	SHOP DRAWINGS, AND TICKETS		----	300/	12.0		
	FIELD MEASURING, 6 PCS		----		---	.75	12.0
	CARTAGE, 5 LOADS		----	2.5	13.0		
	BALANCING (18) OUTLETS, RTU, PRE		----				8.0
	JOB TOTALS	$	12,857		97.0		242.0
	SUB-CONTRACTORS				HR		HR
	REFRIGERATION PIPING, ETC.						
	TEMPERATURE CONTROL						
	INSULATION						

Bid Recap and Markup Sheet

Job **IBM OFFICES** _____ Due Date _____

Location _____ Estimator _____

	HOURS	WAGE RATE	COST		
Shop Labor	97	37.00	$ 3 589		
Field Labor	242	37.00	$ 8 954		
Wage Increase Shop	150	1.00	$ 150		
Wage Increase Field			$		
Overtime			$		
Travel Costs			$		
			$		
			$		
TOTAL LABOR				$	12 693
Raw Materials			$ 2 187		
Equipment			$ 11 868		
			$ 14 055		
Sales Tax 7%			$ 790		
TOTAL MATERIAL AND EQUIPMENT				$	14 854
Subcontracts Insulation, FAI, 200 SQ Ft, 1 1/2"			$ 400		
Refrigeration, Condnr, Pipe insul.			$ 10 500		
Temp. Control			$ 1 500		
			$		
			$		
			$ 12 400		
TOTAL SUBCONTRACTS				$	12 400
TOTAL DIRECT COSTS				$	39 938
Overhead On Labor		30%	$ 3 808		
Overhead On Material and Equipment:		10%	$ 1 485		
Overhead On Subcontractors:		5%	$ 742		
TOTAL OVERHEAD				$	6 035
(% of Total Direct Costs) (% of Sales)					
TOTAL DIRECT AND INDIRECT COSTS				$	45 973
Profit: 5 % of Total Costs x 41,239			$ 2 299		
Performance Bond: % of Total Bid			$		
Financing Costs: Amount $; %			$		
Total O&P Markup $, Percent of Sales %				$	2 299
TOTAL BID PRICE				$	48 272

Budget Check: ___$48,272 ÷ 6,800 = $7.09/sq ft___

_____48,272 ÷ 20 Tons = 2.414_____

Calculating Labor Costs Per Hour

Location __Typical U.S._____ Date _____

☐ Union; Local No._____ Contract Expiration_____ Non Union _____

FRINGE BENEFITS PER HOUR

	Journeyman			Non Union
Base Rate	$26.00	$	$	$12.00
Welfare (Medical) (%)	3.90			1.50
Pension (%)	4.40			.35
Apprentice Fund	.55			- - -
National Training Fund	.13			- - -
Vacation Savings	1.00			.50
Industry Fund	.19			.50
TOTAL BENEFITS %	$10.17	$	$	$ 2.85
TOTAL WITH BASE	$36.17	$	$	$14.85

PAYROLL TAXES AND INSURANCE ✳

F.I.C.A.	7.65%	1.99			.91
Workman's Comp.	8.00%	2.07			.96
Federal Unemployment	.60%	.16			.07
State Unemployment	2.70%	.70			.32
Liability Insurance	1.25%	.33			.15
Property Insurance	1.00%	.26			.12
Association Due	.60%	.16			.07
TOTAL TAXES & INS.	21.80%	5.67			2.60
TOTAL BASE, BENEFITS, TAXES, INS		41.84			17.45

✳ *Percentage of Base Rate*

COST PER POUND BREAKDOWN

	AT $41.84 Base, Union		AT $17.45 Base, Non	
	LBS/HR	COST/LB*	LBS/HR	COST/LB*
Material	44	.43		.43
Shop Labor	25	.95		.40
Field labor	200	1.67		.70
Shop Drawings	800	.20		.09
Cartage		.05		.01
TOTAL DIRECT COSTS		$3.33		$1.63
Indirect Overhead %	30	1.00		1.00
TOTAL COSTS		4.33		2.63
Profit %	5	.22		.13
TOTAL SELL		$4.55		$2.76

*Based on total labor cost above $

Telephone Quotation

Job	IBM Offices

Supplier		Phone	
		By	

QTY	MFGR	DESCRIPTION	ACCESS-ORIES	AMOUNT	
				Each	Total
1	Carrier	Air Handling Unit			$9,953
		20 Tons, 8000 CFM			
		DX, Gas, 2 Stage			
		Grand Total			

NOT INCLUDED				
	Vibration Isolators			

Meets plans and specs ☒	Taxes included ☐
Addendums included ☒	Freight included ☒
Type materials correct ☒	Lead time required 6 weeks
	Price good for 60 days

Bidding Record

Job <u>IBM Office Building</u>   Due Date <u>July 15, 1997</u>

Location _____   Time <u>2 pm.</u>

BID SUBMITTED TO:

Company	Name	Phone	Amount	Remarks
Kemper G.C.	T. Richman	981-2038	$48,272	
Cochran Builders	F. Andrew	439-9252	$48,272	
Quim Contr's	R. Burke	870-8662	$47,050	Exclude Insul.

INCLUSIONS	EXCLUSIONS
Equipment	Power Wiring
Sheet Metal	Painting
Piping	Structural Steel Openings
Insulation	Plumbing
Temp. Control	

ADDENDUMS	ALTERNATES
No. 1 July 8, 1997	None

Drawings Included <u>M1 - M3</u>

Specifications Included <u>Section 15</u>

Remarks _____

This page intentionally left blank

Section II

Budget Estimating

This page intentionally left blank

Chapter 4

Budget Estimating

There Are Three Basic Types of Estimates
1. Conceptual or budget estimating
2. Semi-detailed budget estimating
3. Detailed estimating

BUDGET ESTIMATES

A conceptual budget estimate is a quick, approximate price without extensive quantity surveys or pricing calculations of the specific of items that comprise a bid.

Contractors use conceptual budget estimates in the following ways:
- To help determine if they should bid a job or not.
- To know approximately what the total price of a project should be before preparing a detailed estimate.
- Quick pricing for customer, general contractor or A/E.
- As a check of your estimate after you're done.
- For budgeting purposes, comparing design costs, feasibility studies.
- As a check on current market pricing.
- For negotiating.
- To check incoming bids and quotations.

There are two basic types of budget estimates, one for budgeting the overall price of the project or trade, and the other, the semi-detailed one, for budgeting individual component parts and totaling for the job.

The conceptual estimate issued for overall pricing, is a rough approximation based on some unit of measurement of the building or system:
— 10,000 sq ft Office Building @ $7.00/sq ft for HVAC = $70,000
— 100 Tons of Air Cond. in a Store @ $1500.00/Ton = $150,000
— 100,000 sq ft High School @ $9.00/sq ft for sheet metal = $900,000

Conceptual pricing should normally not be used for bidding with any firm commitment on the price. Your price may be reasonably accurate, even right on, but it also can be off the mark 20 to 25% one way or the other due to variations in design, scope, etc.

You may look at an office building, figure $7.00/sq ft, find out later, upon closer inspection or doing the job, that this particular office building runs $8.00/sq ft and you lose $20,000 on a 20,000 sq ft job.

SEMI-DETAILED SCOPE BUDGET ESTIMATES

Semi-detailed budget estimates, the second basic approach to pricing construction work is far more accurate than a conceptual one. It is "on" more frequently and only has a potential inaccuracy of 10 to 15% while still being a very speedy way to price without the tedious, length takeoffs and extensions.

In semi-detailed estimating you break the job down into component parts, as you would a fully detailed one, listing all the individual items on a summary and extension sheet. But instead of a long exact take off on each and extensive pricing, the quantities and the prices of the various components are budgeted; i.e., they are approximated in some expedient manner.

A thorough preliminary survey is made, again, as you would on a fully detailed bid, to insure covering and knowing the entire scope of the project and so that spec requirements are met.

Ductwork Budgeting
Ductwork weight is approximated by using an average weight per square foot of building, or by measuring the total linear footage with a measuring wheel, and multiplying it times the approximate average weight per linear foot.

Ductwork costs for the project are then found by multiplying the budget price per pound or per foot, based on inspection and judgment, times the approximate weight or linear footage.

Piping Budgeting
- Main Piping Runs Per Foot
- Piping Assembly Hookups To Equipment Per Assembly Price

Equipment Pricing
Equipment costs are determined in one of two ways. If the equipment has been sized and specified suppliers can furnish approximate pricing. The second pricing is budgeting by the cost per CFK square foot, linear foot, etc. or by the size. Equipment labor is determined either by the average size or the specific sizes.

Again, as with conceptual budget estimates, it is very risky business bidding firm quotes based on semi-detailed estimates. However, you can chance it, when the situation demands it by adding 10 or 15% on top to cover the possible range of error.

DETAILED ESTIMATES

The only truly reliable estimate is the fully detailed one which absolutely minimizes the risks of being far too low or high, where the odds are greatly in your favor that the estimate is a true representation of the job.

In a detailed estimate the systems and work are broken down into the smallest type components and sizes in which they can be priced with reasonable accuracy and practicality. All items, are components, are taken off and priced separately by specific sizes, types, accessories, etc.

Budget Estimating HVAC Costs and Engineering Loads Per Square Foot of Building and Per Ton

TYPE BUILDING		HEATING COOLING LOAD		DUCT LOAD	SUPPLY	WEIGHT	TOTAL HVAC SELLING PRICE	
		Btu Per Sq Ft	Sq Ft Per Ton	Btu Per Sq Ft	CFM Per Sq Ft	lbs Per Sq Ft	Per Sq Ft	Per Ton
Apartments, condominiums		25	480	26	0.8	0.2	$5.22	$2,491
Auditoriums		40	300	40	1.3	0.8	13.56	4,066
Banks		48	250	26	1.6	1.2	14.71	3,672
Bowling Alleys		40	300	40	1.3	0.6	10.66	3,197
Churches		36	330	36	1.2	0.5	11.65	3,848
Clubhouses		50	240	40	1.6	1.0	10.39	2,491
Cocktail Lounges		70	170	30	1.3	1.2	14.71	2,497
Computer Rooms		140	85	20	4.5	1.5	30.67	2,602
Colleges:	Admin., Classrooms	44	270	42	1.5	1.3	21.50	5,810
	Dormitories	--	--	25	--	0.3	12.28	--
	Gyms, Fieldhouses	--	--	40	0.5	0.3	9.12	--
	Science Bldgs.	54	220	55	1.8	1.4	16.47	3,622
Court Houses		50	240	36	1.6	1.2	14.83	3,555
Fire Stations		--	--	35	--	0.7	9.12	--
Funeral Homes		30	400	32	0.9	0.8	10.36	3,522
Hospitals		44	170	28	1.4	1.5	28.76	4,885
Hotels		34	350	28	1.1	0.5	7.99	3,432
Housing for Elderly --		--	28	--	0.2	7.84	--	
Jails		25	480	30	0.8	1.2	13.93	5,810
Laboratories		60	200	50	2.0	1.5	27.72	5,546
Libraries		46	260	37	1.5	1.3	15.95-	4,148
Manufacturing Plants		40	300	36	1.3	0.4	5.85	1,748
Medical Centers, Clinics		35	340	35	1.1	1.0	10.36	3,522
Motels		30	400	30	1.0	0.s	6.72	2,683
Municipal Bldgs., Town Halls		45	265	36	1.4	1.2	15.21	4,027
Museums		34	350	40	1.1	0.8	11.41	3,990
Nursing Homes		43	280	34	1.4	0.5	13.69	3,827
Office Bldgs:	Low Rise	35	340	35	1.1	.5 to .8	10.36	3,522
	High Rise	40	300	30	1.3	1.1	14.83	4,446
	Small Plant Off.	34	350	35	1.1	0.5	7.40	2,590
Police Stations		42	285	35	1.3	1.0	9.99	2,850
Post offices		44	270	40	1.4	1.2	15.21	4,103
Project Homes		--	--	28	--	0.1	3.64	--
Restaurants		80	150	35	2.0	1.0	16.34	2,224
Residences		25	500	30*	0.8	0.3	4.84	2,406
Schools:	Elementary	--	--	26	0.8	0.7	11.41	--
	Middle, Jr. Highs	36	333	26	1.0	1.1	14.56	4,409
	High schools	33	360	40	1.1	1.3	13.44	4,831
	Vocational	20	600	40	0.9	1.5	30-90	--
Stores:	Beauty Shops	63	190	30	2.0	1.3	16.47	3,127
	Department	34	350	30	1.1	0.7	9.26	3,235
	Discount	30	400	32	1.1	0.4	6.36	2,217
	Retail Shops	50	340	32	1.6	0.8	10.64	3,616
	Shopping Centers	30	400	30	1.1	0.4	5.22	2,076
	Supermarkets	30	400	32	1.0	0.4	5.45	2,179
Theaters		40	300	40	1.3	0.8	10.51	3,152
Warehouses		--	--	20	--	0.2	3.82	--

1. Cooling load based on is temp. differences, 50% RH, 400 CFM per ton.
2. Heating load based on 70 temp. differences and is the output Btu.
    Apply inefficiency factor to heating equipment for input Btu.
3. Price includes all HVAC material, labor and subs and overhead and profit.
    *R-11 wall and R-19 ceiling insulation

Budget Estimating Galvanized Ductwork
Per Pound and Per Foot
Standard Low Pressure HVAC Rectangular Galvanized
25 Percent Fittings, New Construction, 10 Foot High, 1st Floor

SIZE	SEMI-PERIM Inches	GAUGE	LB/FT w/20% Waste	SQ FT/FT No Waste	SELLING PRICE * Furnished & Installed Per Lb.	Per Ft.
6x6	12	26 Ga.	2.8	2.0	$5.14	$14.39
12x6	18		3.3	3.0	4.96	16.34
12x12	24		4.4	4.0	4.77	22.07
18x6	24	24 Ga.	5.6	4.5	4.77	26.71
18x12	30		7.0	5.0	4.62	33.24
24x9	33		7.7	6.5	4.54	34.94
24x12	36		8.4	6.0	4.47	37.49
24x15	39		9.2	6.5	4.47	41.07
30x12	42		9.8	7.0	4.38	42.99
30x18	48		11.2	8.0	4.31	48.26
30x24	54		12.6	9.0	4.23	53.34
36x12	48	22 Ga.	13.6	8.0	4.23	57.57
36x18	54		15.3	9.0	4.13	63.11
36x24	60		17.0	10.0	4.00	68.04
42x12	54		15.3	9.0	4.13	63.11
42x18	60		17.0	10.0	4.00	68.04
42x24	66		18.7	11.0	3.97	74.27
48x12	60		17.0	10.0	4.00	68.04
48x18	66		18.7	11.0	3.97	74.27
48x24	72		10.4	12.0	3.94	80.41
54x24	78		22.1	13.0	3.91	86.40
54x30	84		23.8	14.0	3.88	92.32
54x36	90		25.5	15.0	3.85	98.12
60x15	78	20 Ga.	26.0	13.0	3.88	100.85
60x24	84		28.0	14.0	8.50	108.60
60x30	90		30.0	15.0	3.85	115.44
72x24	96		32.0	16.0	3.82	122.14
72x30	102		34.0	17.0	3.79	128.75
72x36	108		35.3	18.0	3.74	132.03
84x30	114		38.0	19.0	3.82	140.38
84x36	120		40.0	20.0	3.67	146.53
84x42	126		42.0	21.0	3.63	152.56
96x24	120	18 Ga.	52.6	21.0	3.60	189.44
96x36	132		57.2	22.0	3.57	204.25
96x48	144		62.4	24.0	3.54	220.90
96x72	168		72.8	28.0	3.51	255.51
96x96	192		83.2	32.0	3.48	289.37

*Selling price based on $.42 per pound for galvanized, $39.00 per hour for labor, a 30 percent markup on direct costs for overhead and a 5 percent markup for profit. The price includes material and all labor for the fabrication, installation, drafting and shipping.

CORRECTION FACTORS ON SELLING PRICE OF GALVANIZED DUCTWORK
1. Percent fittings by weight,

| 15% | ....0.09 | 25% | ....1.00 |
| 35% | ....1.08 | 45% | ....1.15 |

2. Different wage rates (incl. base pay, fringes, payroll taxes, ins.)

$12.00 (per hour)	.... 0.66	$22.00	.................0.89
15.00	.........................0.77	26.50	.................1.00
18.00	..........................0.80	30.00	.................1.08

3. Overall residential factor for 15 percent fittings, $12.00 per hour gross wages and lighter gauges ... 0.50 times cost per lb or per foot.

Example: 24" x 8 $4.15/lb equals $2.08/lb
$31.90/ft equals $15.95/ft

Installed Price Per Square Foot for
Different Types of Ductwork
Average Size 24"x12" in a Typical Mix
25 Percent Fittings by Square Feet

HVAC DUCTWORK

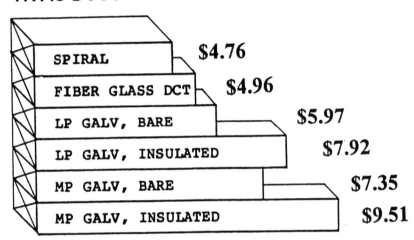

SPIRAL	$4.76
FIBER GLASS DCT	$4.96
LP GALV, BARE	$5.97
LP GALV, INSULATED	$7.92
MP GALV, BARE	$7.35
MP GALV, INSULATED	$9.51

INDUSTRIAL DUCTWORK

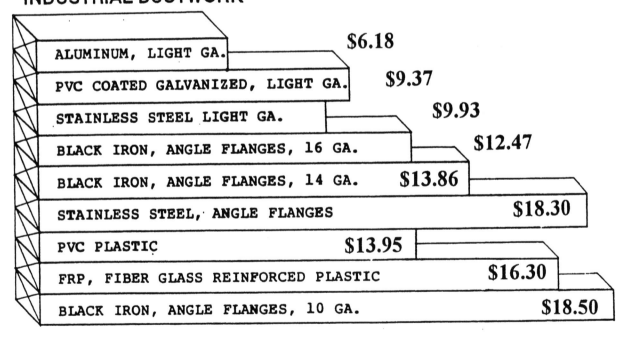

ALUMINUM, LIGHT GA.	$6.18
PVC COATED GALVANIZED, LIGHT GA.	$9.37
STAINLESS STEEL LIGHT GA.	$9.93
BLACK IRON, ANGLE FLANGES, 16 GA.	$12.47
BLACK IRON, ANGLE FLANGES, 14 GA.	$13.86
STAINLESS STEEL, ANGLE FLANGES	$18.30
PVC PLASTIC	$13.95
FRP, FIBER GLASS REINFORCED PLASTIC	$16.30
BLACK IRON, ANGLE FLANGES, 10 GA.	$18.50

Installed price per square foot includes material, shop labor, field labor, shop drawings, shipping and a 35 percent markup on costs for overhead and profit. Labor is based on $39.00 per hour.

Per Square Foot Ductwork Cost Comparison
Average Size 24"x12"
25 Percent Fittings By Square Feet

TYPE DUCT	DESCRIPTION	DIRECT MATERIAL COST* per sq ft		SHOP LABOR		FIELD LABOR		BUDGET PRICE with 30% O&P per sq ft
				per ft per hr	hr per sq ft	per ft per hr	hr per sq ft	
STAINLESS STEEL	24 gauge with cleats $3.18	30	0.03	17	0.06		$8.81	
	18 gauge welded seams metal flanged and bolted	6.50		10	0.10	15	0.07	16.90
ALUMINUM	20 gauge .032" thick cleat connections	0.87		33	0.03	19	0.05	5.34
FRP	3/16" thick coupling or flange connection 1-3.32	4.40		12	0.08	is	0.07	
PVC	1/8" or 3/16" thick flange or slip connections	5.12		23	0.04	16	0.06	12.03
PVC (PVS) COATED GALVANIZED	22 gauge with cleats or flanged and bolted	2.80		30	0.03	17	0.06	8.31
BLACK IRON	metal flanges, bolted, 10 gauge	3.27		9	0.11	10	0.10	14.96
	14 gauge	1.82		11	0.09	14	0.07	10.60
	angle flanges, bolted 10 gauge	6.67		8	0.13	9	0.11	20.65
	14 gauge	2.29		10	0.10	13	0.08	11.95
	16 gauge, kitchen exhaust welded connections, NFPA 96	1.42		16	0.06	10	0.10	10.09
FIBER GLASS	800 board, stapled and taped, rod reinforcing	1.49		55	0.02	30	0.03	4.56
GALVANIZED	low pressure, unlined	0.74		38	0.03	22	0.05	4.60
	low pressure, 1" lining, 1-1/2 lb.	1.47		22	0.05	21	0.05	6.62
SPIRAL	pipe and fittings, medium pressure, slip connections sealed	0.75		47	0.02	24	0.04	4.17

Sell budget prices include material, shop and field labor, shop drawings, field measuring, shipping, and a 30% markup for overhead and profit.

*Material costs includes sheet metal, connection materials, hangers, scrap hardware.

Labor rate $39.00 per hour.

Budget Estimating Insulation and Lining Per Foot
Standard HVAC Rectangular Ductwork
25 Percent Fittings by Square Footage

SIZE	SEMI-PERIM (w & d) Inches	SQ FT/FT with 15% Waste	SELLING PRICE Furnished & Installed per Ft	
			Insulation	Lining
6x6	12	2.3	$5.06	$4.32
12x6	18	3.5	7.70	5.64
12x12	24	4.6	10.12	8.63
18X9	24	5.2	11.46	9.75
18x12	30	5.8	12.76	10.79
24x9	33	7.5	16.52	14.09
24x12	36	6.9	15.21	12.95
24x15	39	7.5	16.52	14.09
30x12	42	8.0	17.61	15.02
30x18	48	9.2	20.26	17.27
30x24	54	10.4	22.88	19.52
36x12	48	9.2	20.26	17.27
36x18	54	10.4	22.88	19.52
36x24	60	11.5	25.32	21.59
42x12	54	10.4	22.88	19.52
42x18	60	11.5	25.32	21.59
42x24	66	12.7	27.96	23.84
48x12	60	11.5	25.32	21.59
48x18	66	12.7	27.96	22.54
48x24	72	13.8	30.37	25.91
54x24	78	15.0	33.02	28.16
54x30	84	16.1	35.43	30.24
54x36	90	17.3	38.08*	32.486
60x18	78	15.0	33.02	28.16
60x24	84	16.1	35.43	30.24
60x30	90	17.3	42.70	32.49
72x24	96	18.4	40.49	33.80
72x30	102	19.6	43.14	36.81
72x36	108	20.7	45.55	38.86
84x30	114	21.9	48.20	39.43
84x36	120	23.0	50.63	43.19
84x42	126	24.1	53.04	45.26
96x24	120	24.2	53.27	45.44
96x36	132	25.3	55.69	68.01
96x48	144	27.6	60.75	51.83
96x72	168	32.2	70.88	60.46
96x96	192	36.8	80.99	69.12

1. Lining is based on 1 inch thick 1-1/2 pound density fiber glass blanket at $1.48 per sq. ft. installed. Installed selling prices based on $39.00 per hour for labor, a 25 percent markup for overhead and a 5 percent markup for profit. Includes all pins, tapes, cement, materials and labor.

2. Galvanized ductwork weight must be increased 12 percent to compensate for increasing the metal duct size one inch for the lining.

Budget Estimating External Insulation Per Foot
Standard HVAC Rectangular Ductwork
25 Percent Fittings By Square Footage

SIZE	SEMI-PERIM -------- (w & d) Inches	SQ FT/FT With 15% Waste	SELLING PRICE Furnished & Installed per ft Insulation
6x6	12	2.3	$5.06
12x6	18	3.5	7.70
12x12	24	4.6	10.12
18x9	24	5.2	11.46
18x12	30	5.8	12.76
24x9	33	7.5	16.52
24x12	36	6.9	15.21
24x15	39	7.5	16.52
30x12	42	8.0	17.61
30x18	48	9.2	20.26
30x24	54	10.4	22.88
36x12	48	9.2	20.26
36x18	54	10.4	22.88
36x24	60	11.5	25.32
42x12	54	10.4	22.88
42x18	60	11.5	25.32
42x24	66	12.7	27.96
48x12	60	11.5	25.32
48x18	66	12.7	27.96
48x24	72	13.8	30.37
54x24	78	15.0	33.02
54x30	84	16.1	35.43
54x36	90	17.3	38.08
60x18	78	15.0	33.02
60x24	84	16.1	35.43
60x30	90	17.3	42.70
72x24	96	18.4	40.49
72x30	102	19.6	43.14
72x36	108	20.7	45.55
84x30	114	21.9	48.20
84x36	120	23.0	50.63
84x42	126	24.1	53.04
96x24	120	24.2	53.27
96x36	132	25.3	55.69
96x48	144	27.6	60.75
96x72	168	32.2	70.88
96x96	192	36.8	80.99

1. External insulation based on 1-1/2 inch thick 3/4 pound density fiber glass with vapor barrier, at $1.72 per sq. ft. installed.

Budget Estimating Insulation
External Wrapping For Ductwork Per Square Foot

ITEM	THICKNESS Inches	DENSITY Lbs Per Cu Ft	LABOR Sq Ft Per Hour	MATERIAL COST Per Sq Ft	SELLING PRICE* w/O&P Per Sq Ft
FIBER GLASS BLANKET, RFK					
Vapor Barrier or Vinyl	1	3/4	50	$0.46	$1.91
	1-1/2	3/4	45	0.62	2.34
	2	3/4	40	0.81	2.85
FIBER GLASS, RIGID BOARD					
With RFK Vapor Barrier	1	3	16	0.53	4.21
With 8 oz. Canvas Cover	1	3	15	0.71	4.77
With 8 oz. Canvas and Painted	1	3	14	0.80	5.18
With RFK Vapor Barrier	2	3	13	0.62	5.11
With 8 oz. Canvas Cover	2	3	12	0.54	5.28
With 8 oz. Canvas and Painted	2	3	11	0.60	5.78
OUTSIDE.FIBER GLASS, RIGID BOARD					
With 2 Layers black mastic, glass mesh	1	4	--	-----	5.09
	2	4	--	-----	5.67
KITCHEN EXHAUSTS & BREECHING					
Calcium Salicate Block (K Block)					
Rectangular	2	--	8	2.03	10.29
Round	2	--	7	2.72	12.55
With 1/2" Cement					
Rectangular	2	--	5	3.61	17.18
Round	2	--	4	4.75	21.94
Fiber Glass, Rigid Board					
with covering and paint	2	4	8	1.33	8.93

*Includes 15% waste, pins, staples, tape, installation, supervision, 20% overhead and 5% profit. Based on union wages of $39.00 per hour, 10 foot high ducts, lower floors, normal space conditions.

CORRECTION FACTORS
    1. Congested ceiling spaces, areas 1.15
    2. Wage rate at $39.00 per hour

Budget Estimating Copper Tubing
Installed Price Per Foot
95/5 Solder, 300F

Includes 2 Fittings Per 20 Feet of Pipe, Hangers and Solder
Standard Installation Conditions, Lower Floors, 10 Ft High

DIA. Inches	L	M	K	DWV
1/2	$9.32	$8.39	$10.03	$8.81
3/4	11.83	10.64	13.23	11.17
1	15.OB	13.26	16.71	13.93
1-1/4	18.97	17.08	23.41	17.95
1-1/2	21.70	19.54	23.77	20.51
2	28.00	25.19	30.71	26.46
2-1/2	38.36	33.68	41.00	35.36
3	45.57	41.03	49.91	43.07
4	65.11	58.59	75.95	61.52
5	117.20	105.48	138.63	110.75
6	154.08	138.68	200.01	145.62
8	282.13	228.53	335.86	239.20

Prices include all material costs, labor at $51.00 per hour and overhead and profit markup.

Budget Estimating Steel Piping
Installed Price Per Foot
Includes 2 Fittings Per 20 Feet of Pipe, Hangers
Standard Installation Conditions, Lower Floors, 10 Ft High

DIA. Inches	SCREWED Sch. 40	WELDED Sch. 40	FLANGED Sch. 40	GROOVED Sch. 40
1/2	$8.54	----	----	----
3/4	9.00	----	----	7.70
1	10.74	13.96	----	8.47
1-1/4	12.24	15.30	----	9.24
1-1/2	13.85	16.49	----	11.54
2	17.85	18.65	33.94	15.24
2-1/2	23.68	24.61	40.17	19.24
3	28.98	29.91	46.18	23.09
4	37.99	39.72	62.34	30.55
5	60.03	55.23	78.50	40.17
6	73.88	66.94	96.20	53.56
8	99.28	83.40	125.44	80.97
10	115.44	95.82	187.01	108.85
12	168.54	129.93	253.97	135.76
14	----	----	295.53	----
16	----	----	341.70	----
18	----	----	369.41	----
20	----	----	420.20	----
22	----	----	434.05	----
24	----	----	458.68	----

Prices include all material costs, labor at $51.00 per hour and overhead and profit markup.

Budget Estimating PVC Pressure Piping
Installed Price Per Foot
Includes 2 Fittings Per 20 Feet of Pipe, Hangers and Solvent
Standard Installation Conditions, Lower Floors, 10 Ft High

DIA	LABOR	TOTAL MATERIAL & LABOR	
	Man Hours	Direct Cost	With 20% O&P
1/2″	0.140	$6.81	$8.17
3/4″	0.148	$7.18	$9.33
1″	0.129	$6.14	$7.98
1-1/4″	0.147	$7.02	$9.13
1-1/2″	0.149	$7.23	$9.40
2″	0.156	$7.75	$10.07
2-1/2″	0.177	$9.83	$12.78
3″	0.229	$12.90	$16.76
4″	0.317	$13.36	$17.37

Direct labor costs are $39.00 per hour.

Budget Estimating Cast Iron Piping, Hub and
Spigot Installed Price Per Foot
Includes 2 Fittings Per 20 Feet of Pipe, Hangers
Standard Installation Conditions, Lower Floors, 10 Ft High

DIA	LABOR	TOTAL MATERIAL & LABOR	
	Man Hours	Direct Cost	With 20% O&P
2″	0.212	$12.64	$15.16
3″	0.305	$18.41	$23.93
4″	0.331	$20.33	$26.43
5″	0.378	$25.27	$32.85
6″	0.432	$29.43	$38.26
8″	0.574	$45.97	$59.76
10″	0.761	$67.39	$87.61
12″	0.835	$87.83	$114.18
15″	1.112	$146.28	$190.16

Direct labor costs are $39.00 per hour.

Threaded Bronze and Iron Valves

SIZE	TYPE	DIRECT MATERIAL COST	LABOR	TOTAL MATERIAL & LABOR	
	Material		Man Hours	Direct Cost	With 30% O&P
GATE					
1/2	Bronze	$22.79	0.65	$48.14	$62.58
3/4	Bronze	29.53	0.72	57.61	74.89
1	Bronze	35.74	0.78	66.16	86.01
1-1/4	Bronze	47.95	0.91	83.44	108.47
1-1/2	Bronze	61.02	0.98	99.24	129.01
2	Bronze	85.00	1.04	125.56	163.22
2-1/2	Iron	283.32	1.80	353.52	459.57
3	Iron	348.69	1.90	422.79	549.62
4	Iron	435.86	2.00	513.86	668.02
5	Iron	0.00	0.00	0.00	0.00
6	Iron	0.00	0.00	0.00	0.00
8	Iron	0.00	0.00	0.00	0.00
CHECK					
1/2	Bronze	$24.42	0.65	$49.77	$64.70
3/4	Bronze	27.65	0.72	55.73	72.44
1	Bronze	36.19	0.78	66.61	86.59
1-1/4	Bronze	51.06	0.91	86.55	112.52
1-1/2	Bronze	57.72	0.98	95.94	124.72
2	Bronze	88.80	1.04	129.36	168.17
2-1/2	Iron	213.12	1.80	283.32	368.32
3	Iron	255.30	1.90	329.40	428.22
4	Iron	388.50	2.00	466.50	606.45
5	Iron	0.00	0	0.00	0.00
6	Iron	0.00	0	0.00	0.00
8	Iron	0.00	0	0.00	0.00
BALL					
1/2	Bronze	$7.87	0.65	$33.22	$43.19
3/4	Bronze	12.55	0.72	40.63	52.82
1	Bronze	16.10	0.78	46.52	60.48
1-1/4	Bronze	28.09	0.91	63.58	82.65
1-1/2	Bronze	35.74	0.98	73.96	96.15
2	Bronze	44.40	1.04	84.96	110.45
2-1/2	Iron	0.00	0.00	0.00	0.00
3	Iron	0.00	0.00	0.00	0.00
4	Iron	0.00	0.00	0.00	0.00
5	Iron	0.00	0	0.00	0.00
6	Iron	0.00	0	0.00	0.00
8	Iron	0.00	0	0.00	0.00

Direct labor costs are $39.00 per hour.

Threaded Bronze and Iron Valves

SIZE	TYPE — Material	DIRECT MATERIAL COST	LABOR Man Hours	TOTAL MATERIAL & LABOR Direct Cost	With 30% O&P
GLOBE					
1/2	Bronze	$32.53.	0.65	$57.88	$75.24
3/4	Bronze	39.86	0.72	67.94	88.32
1	Bronze	66.60	0.78	97.02	126.13
1-1/4	Bronze	95.46	0.91	130.95	170.24
1-1/2	Bronze	117.66	0.98	155.88	202.64
2	Bronze	179.82	1.04	220.38	286.49
2-1/2	Iron	543.90	1.80	614.10	798.33
3	Iron	610.50	1.90	684.60	889.98~
4	Iron	943.50	2.00	1021.50	1327.95
5	Iron	0.00	0	0.00	0.00
6	Iron	0.00	0	0.00	0.00
8	Iron	0.00	0	0.00	0.00
ANGLE					
1/2	Bronze	$44.40	0.65	$69.75	$90.68
3/4	Bronze	57.72	0.72	85.80	111.54
1	Bronze	82.14	0.78	112.56	146.33
1-1/4	Bronze	108.78	0.91	144.27	187.55
1-1/2	Bronze	144.30	0.98	182.52	237.28
2	Bronze	222.00	1.04	262.56	341.33
2-1/2	Iron	0.00	0.00	0.00	0.00
3	Iron	0.00	0.00	0.00	0.00
4	Iron	0.00	0.00	0.00	0.00
5	Iron	0.00	0	0.00	0.00
6	Iron	0.00	0	0.00	0.00
8	Iron	0.00	0	0.00	0.00

Direct labor costs are $39.00 per hour.

Budget Control Pricing For Built-Up Systems In New Buildings

	PERCENT OF EQUIPMENT COSTS
Under $25,000	24%
$25,000 to $50,000	20
$50,000 to $100,000	17
$100,000 to $250,000	15
$250,000 to $500,000	12
Add for computerized controls	10%

CONTROLS FOR PACKAGE SYSTEMS

Under $25,000	8%
$25,000 to $100,000	7
$100,000 to $500,000	6

Pneumatic Temperature Control
Compressor Systems
Includes Compressor, Tank, Tubing, Dryer, Filter, PRV Valve, Motor, Starter

HP	DIRECT MATERIAL COST			LABOR	TOTAL MATERIAL & LABOR	
	Duplex Compress	Other Items	Total Matl Costs	Man Hours	Direct Cost	With 30% O&P
1/2	$1,140	$4,484	$5,624	20	$6,404	$8,325
3/4	1,332	5,920	7,252	21	8,071	10,492
1	1,532	6,090	7,622	23	8,519	~11,075
1-1/2	1,877	6,559	8,436	25	9,411	12,234
2	2,168	7,452	9,620	30	10,790	14,027
3	2,512	8,440	10,952	35	12,317	16,012
5	2,815	10,505	13,320	40	14,880	19,344

1. Includes installation of compressor, piping hook up, valves, etc.
2. To furnish and install 3/81, diameter copper trunk line tubing add $6.50 per foot
3. If poly vinyl tubing is used instead of copper tubing deduct 25% from copper tubing.

Direct labor costs are $39.00 per hour.

Two Way Control Valves

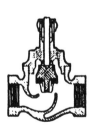

DIA. Inches	Cv	DIRECT COST Each	TYPE CONNECT & PRESS	LABOR Man Hours	TOTAL MATERIAL & LABOR	
					Direct Cost	With 30% O&P
1/2	.63,1.0, 1.6	$205.72	Screw, hp	0.55	$227	$295
1/2	2.5, 4.0	127.28	Screw	0.55	149	193
3/4	6.3	214.60	Screw, hp	0.60	238	309
3/4	6.3	131.72	Screw	0.60	155	202
1	10.0	251.60	Screw, hp	0.65	277	360
1	10.0	140.60	Screw	0.65	166	216
1-1/4	16.0	287.12	Screw, hp	0.77	317	412
1-1/4	16.0	173.16	Screw	0.77	203	264
1-1/2	25.0	340.40	Screw, hp	0.83	373	485
1-1/2	25.0	230.88	Screw	0.83	263	342
2	40.0	313.76	Screw, hp	0.88	348	453
2	40.0	297.48	Screw	0.88	332	431
2-1/2	63.0	411.44	Screw	2.00	489	636
2-1/2	63.0	830.28	Flange	2.00	908	1,181
3	100.0	555.00	Screw	2.10	637	828
3	100.0	994.56	Flange	2.10	1,076	1,399
4	160.0	1376.40	Flange	2.40	1,470	1,911

NOT INCLUDED:
1. Control motor
2. Linkages
3. Plug in balance relays

hp: High pressure

Direct labor costs are $39.00 per hour.

Three Way Control Valves

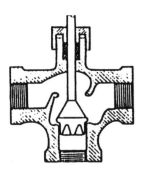

MIXING

DIA.		DIRECT COST	TYPE CONNECT & PRESS	LABOR	TOTAL MATERIAL & LABOR	
	Cv					
Inches		Each		Man Hours	Direct Cost	With 30% O&P
1/2	4.0	$159.84	Screw	0.55	$181	$236
3/4	6.3	176.12	Screw	0.60	200	259
1	10.0	205.72	Screw	0.65	231	300
1-1/4	16.0	241.24	Screw	0.77	271	353
1-1/2	25.0	284.16	Screw	0.83	317	411
2	40.0	334.48	Screw	0.88	369	479
2-1/2	63.0	907.24	Flange	2.00	985	1,281
3	100.0	1073.00	Flange	2.10	1,1S5	1,501
4	160.0	1796.72	Flange	2.40	1,890	2,457

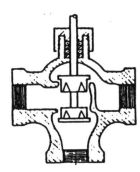

DIVERTING

DIA.		DIRECT COST	TYPE CONNECT & PRESS	LABOR	TOTAL MATERIAL & LABOR	
	Cv					
Inches		Each		Man Hours	Direct Cost	With 30% O&P
2-1/2	63.0	1182.52	Flange	2.00	1,261	1,639
3	100.0	1324.60	Flange	2.10	1,407	1,828

NOT INCLUDED:
1. Control motor
2. Linkages
3. Plug in balance relays

hp: High pressure

Direct labor costs are $39.00 per hour.

Energy Management Systems
Indirect, Low Voltage, Hardwire

EQUIPMENT		PRICES	
	10	Channel EMS unit W7010H*	$4,603
	20	Channel EMS unit W7020H*	$7,548
INSTALLED PRICES			
	10	Channel @ $1,200/pt	$17,760
	20	Channel @ $1,000/pt	$29,600

COMMUNICATIONS LINK HARDWARE $1,776

REMOTE COMPUTER
For Programming and Monitoring $2,960 to $4,440
(Can put whole program on a floppy disk)

FUNCTIONS
- Time of Day Scheduling
- Optimum Start/Stop
- Demand Limiting
- Duty Cycling, Temperature Compensated by Temperatures in spaces
- Monitoring Outside Air and Indoor Air Temperatures

*Approximate Honeywell Contractor Price
Direct labor costs are $39.00 per hour.

Budget Estimating Electrical Work Motors
Induct Motors, 1800 RPM, 230/460 V,
3 Phase, 60 Cycle, T Frame

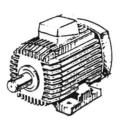

HP	DIRECT COSTS OF MOTOR			LABOR	WIRING HOOKUP	
	Totally DRIPPROOF	Explosion Enclosed Fan Cooled	Install Proof	Labor Motor Hours	Direct Hours	Material Cost
1/4	$80		$90	1.4	0.5	$92
1/2	136	145		1.6	0.5	92
1	219	256	542	2.0	0.6	92
1-1/2	243	283	570	2.2	0.6	92
2	262	311	591	2.5	0.6	92
3	277	345	679	2.8	0.7	92
5	311	401	832	3.0	0.8	102
7-1/2	474	571	1,137	4.0	1.0	111
10	571	690	1,341	5.0	1.3	111
is	767	922		6.0	1.5	138
20	959	1,159		7.0	1.5	179
25	1,137	1,406		8.0	1.5	203
30	1,332	1,658		9.5	2.2	203
40	1,672	2,210		12.0	3.8,	280
so	2,026	2,809		15.0	4.2	318
60	2,565			18.0	4.2	369
75	3,107			21.0	4.8	434
100	3,993			24.0	5.0	490

CORRECTION FACTORS
1. Costs of "U" frame motors: dripproof and TEFC   1.3
                                     explosion proof   1.15

Budget Estimating Starters
Solid State Soft Motor Starters 230/460 Volt, 3 Phase

HP	STARTER	SIZE MATERIAL COST		DIRECT LABOR	TOTAL MATERIAL & LABOR	
		Each	Per HP	Man Hours	Direct Cost	With 30% O&P
5	0	$1,080	$216.08	2.8	$1,190	$1,546
10	1	1,184	$118.40	3.1	1,305	1,696
15	2	1,317	$87.81	3.5	1,454	1,890
20	2	1,421	$71.04	3.5	1,557	2,024
25	2	1,524	$60.98	3.5	1,661	2,159
30	3	1,628	$54.27	4.5	1,804	2,345
40	3	1,924	$48.10	4.5	2,100	2,729
50	4	2,220	$44.40	5.3	2,427	3,155
60	4	2,405	$40.08	5.3	2,612	3,395
75	4	3,330	$44.40	5.3	3,537	4,598
100	5	4,070	$40.70	6.5	4,324	5,621
125		4,810	$38.48		4,810	6,253
150		5,550	$37.00		5,550	7,215
200		7,215	$36.08		7,215	9,380
250		8,880	$35.52		8,880	11,544

Starter

1. Combination magnetic, fusible disconnect, HOA, pushbutton, transformer, pilot light, MEMA 1 enclosure.

Size	Direct Cost	Total Install & Hookup	Price W/Hookup O & P
00	$300	2.5	$570
0	395	2.8	697
1	488	3.1	925
2	586	3.5	1,091
3	1,024	4.5	1,757
4	1,902	5.3	2,892

2. Manual fractional horsepower starter with overload heaters.........$20.20
3. Manual starter I to 10 HP, overload, start-stop buttons, HOA........$85.47

Budget Estimating Wiring
Per Linear Foot For 3 Wires
Furnished And Installed 12 Feet High

AMPs	Wire Stranded THW, Copper		Conduit Rigid Galv.		Labor HRS/LF Conduit & 3 Wires	Total Installed Price With 30* O&P
	Size	3 Wires	Cost Dia.	Cost LF		
15	#14	$0.22	1/2"	$1.10	0.070	$6.35
20	12	0.30	1/2	1.10	0.074	6.71
30	10	0.44	1/2	1.10	0.082	7.44
45	8	0.77	3/4	1.49	0.100	9.57
65	6	1.01	1	2.07	0.120	11.96
85	4	1.51	1	2.07	0.130	13.28
100	3	1.89	1-1/4	2.71	0.146	15.66
115	2	2.26	1-1/4	2.71	0.149	16.34
130	1	3.20	1-1/4	2.71	0.152	17.75
150	1/0	3.89,	1-1/2	3.20	0.184	21.42
175	2/0	4.82	1-1/2	3.20	0.194	23.29
200	3/0	5.93	2	4.37	0.204	26.92

Labor at $51.00 per hour; Selling price includes 30% markup on material.
1. Thin wall conduit versus rigid installed price............95

Motor And Starter Engineering Data
3 Phase, 1800 RPM, Induction 230/460V

HP	Full Load Amps			Starter Size		Percent Efficiency
	115V	230V	460V	230V	460V	
1/2	4	2	1.0	00	00	70
3/4	5.6	2.8	1.4	00	00	72
1	7.2	3.6	1.8	0	00	79
1-1/2	10.4	5.2	2.6	0	0	80
2	13.6	6.8	3.4	0	0	80
3	19.2	9.6	4.8	0	0	81
5	30.4	15.2	7.6	0	0	83
7-1/2		22	11	1	0	85
10		28	14	1	1	85
15		42	21	2	1	86
20		54	27	2	1	87
25		68	34	2	2	88
30		80	40	3	2	89
40		104	52	3	2	89
so		130	65	4	3	89
60			77	5	4	89
75			96	5	4	90
100		248	124	5	4	90

RULE OF THUMB ON AMPS
1. Lower voltages draw higher amps. A 115V motor draws half the amps of a 230V.
2. Higher voltages draw lower amps. A 230V motor draws double a 460V.
3. Single phase motors draw double the amps of 3 phase motors.
4. Rough calculation of amps for higher horse powers.
    Amps = 2.64 x horse power for 230 volts
    Amps = 1.32 x horse power for 460 volts

General Construction Equipment Rental

Abrasive Hole Cutting Saws
Rents for $52/day or $130/week.

Backhoes
Backhoes for excavating rent for $265/day or $800/week.

Structural Support On Roof For HVAC Equipment
For a 15-foot span between 2 joists or beams, an 8 inch "I" beam, running 15.3 lbs
per linear foot will safely support a small or moderate siz6 air conditioner,
fan or condenser.

Concrete Pads
SMALL PADS, 4 x 4 feet, 6" high, 1/2 cubic yard
     Singles $480 minimum
     Multiples $410/cubic yard

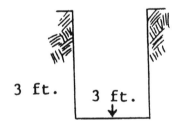

LARGE PADS, 20 x 10 feet, 6" high, 3 cubic yards
     Singles $480 minimum
     Multiples $350/cubic yard

EXCAVATING AND BACKFELLING
3 ft x 4 ft deep trench, machine excavated backfilled and tamped, runs around
$28.00/cubic yard for regular soil.

Sandy soil is 25% less, clay soil is 50% more.

Removal Work
Ductwork removal is about 50%, in existing buildings, of what the new installation rate is.

Equipment removal is 33% of new installation rate.

Cranes and Lifting Equipment Rental

Telescoping Hydro Cranes
Range in capabilities between 4 and 30 tons. 15-ton unit lifts 3000 lbs 70 feet high at 30°, 100 high with a job.

15 ton crane most commonly used .... 1/2 day $540
Full day $750

Crawler Cranes, Lattice Type
Have to be assembled on the job site. Takes 3 men to assemble and 3 to disassemble.

Total cost runs the price of the labor plus $830 to $1,015 per day rental.

Linden Tower Cranes
Counter balanced, rotating crane raised up with the building as it is erected. It sits in the middle and swings around as needed.

Common rental fee from general contractor is $182 to $216 per hour.

Tower Hoist Rental $90 per hour

Helicopters
Helicopter that will lift about 5000 pounds; may run around $1,250 base fee plus $126 per lift Takes about half an hour to sling and hoist each piece as an average.

Scissors Hoist
Manually propelled scissors hoist, 18 foot platform height, 2-1/2 x 6-1/2 foot platform.
Rent $414/week $2,000/month

Self propelled, gas powered scissors hoist, 2000 lb capacity, 5'8" x 13'5" platform.
Platform height is 25 feet.
Rent $773/wee $2,300/month

Forklift Trucks
Truck with 15 foot lift rents for $83 per day.

This page intentionally left blank

Section III

Equipment Estimating

This page intentionally left blank

Chapter 5

Heating and Cooling Equipment

The equipment tables in Section III of this manual are designed to maximize the accuracy of the selections from the tables.

- The left columns in the tables show the size or capacity of the particular piece of equipment, alternate units of measurements if required and sometimes secondary factors-which altogether make selections more specific and realistic.

   Column one might be the capacity of the equipment in typical units such as GPM, BtuK KWs, watts, MBH, tons of cooling, CFM of air, etc. The size might also be denoted in actual dimensional units such as 24" x 24."

   The second factor in the sizing and capacity columns might involve a condition such as feet of head along with gpm as with pumps.

   In some cases, there may be another condition listed long with gpm and feet of head; such as the horsepower of a motor for a particular unit.

- The middle columns involve the direct material costs for each size plus a unit cost based on the units of measurement being used for the size category. For example, a certain GPM pump might have a unit material cost of $41.00 per gallon.

- The right columns show the labor in man hours, the total direct costs of labor and material together and the installed selling price with overhead and profit included.

Chilled And Hot Water Pumps
Heavy Duty Cast Iron, 3500 RPM

	APPROX			DIRECT MATERIAL COST		LABOR	TOTAL MATERIAL & LABOR	
GPM	HEAD FT	HP	Each	Man Per Gal	Direct Hours	With Cost	30W O&P
20	40	1/2	$480	$23.98	2	$558	$725
40	50	1	818	20.46	3	935	1,216
60	60	2	1,132	18.87	4	1,288	1,675
80	70	3	1,222	15.28	5	1,417	1,843
100	80	5	1,311	13.11	6	1,545	2,009
200	80	7-1/2	2,102	10.51	8	2,414	3,138
300	50	10	2,845	9.48	10	3,235	4,205
400	90	15	3,071	7.68	12	3,539	4,601
500	100	20	3,283	6.57	14	3,829	4,977
600	100	25	3,491	5.82	16	4,115	5,350
800	110	30	4,088	5.11	18	4,790	6,227
1,000	120	30	4,687	4.69	20	5,467	7,107
1,200	120	50	5,090	4.24	22	5,948	7,732
1,400	120	50	5,171	3.69	24	6,107	7,939

CORRECTION FACTORS

1. 1,750 RPM pumps, material   1.50
2. All bronze body, material   1.50

Direct labor costs are $39.00 per hour.

Hot Water Reheat Coils
Flanged, 2 Row
Water 200°F In, 180°F Out, 3-4 FPS
Air 70°F In, 110°F Out, 700 FPM

BtuH	TYPICAL SIZE	SQ FT	GPM	CFM	DIRECT MATERIAL COST		LABOR	TOTAL MATERIAL & LABOR	
		Area			Each	Per Sq Ft	Man Hours	Direct Costs	With 30% O&P
15,100	12x6	0.5	1.5	350	$201	$402.56	1.7	$268	$348
30,200	12x12	1.0	3.0	700	234	233.84	1.8	304	395
45,400	18x12	1.5	4.5	1,050	263	175.63	2.0	341	444
60,500	24x12	2.0	6.1	1,400	290	145.04	2.2	376	489
75,600	24x18	2.5	7.6	1,750	318	127.28	2.4	412	535
90,700	36x12	3.0	9.1	2,100	340	113.47	2.6	442	574
105,800	36x14	3.5	10.6	2,450	370	105.71	2.8	479	623
121,000	36x16	4.0	12.1	2,800	400	99.90	3.0	517	672
151,200	36x20	5.0	15.1	3,500	488	97.68	3.5	625	812
181,400	48x18	6.0	18.1	4,200	592	98.67	4.0	748	972
241,800	48x24	8.0	24.1	5,600	770	96.20	5.0	965	1254

Includes complete installation of coil in duct.

Does not include valves or piping connections.

CORRECTION FACTORS

1. 6 Row Coils Material add 25%, Labor add 15%

Electric Duct Heaters, Slip In
Includes Thermal Cutouts,
Fused Transformers, Contractors
Air Flow Switch, Controls, Fan Interlock
Temperature Rise 50°F

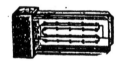

KW	BtuH	SIZE	CFM @ 500 FPM	AMPS	DIRECT MATERIAL COSTS	TOTAL LABOR Man Hours	MATERIAL & LABOR Direct Costs	With 30% O&P
120 VOLT, 1 PHASE, SINGLE STAGE								
1	3,416	8x6	80	9	$253	1.3	$304	$395
2	6,832	8x8	160	17	277	1.3	327	426
3	10,248	10x8	240	25	284	1.3	335	435
4	13,664	10x10	320	34	293	1.4	348	452
5	17,080	12x10	400	42	306	1.4	361	469
208 VOLT, 1 PHASE								
1	3,416	8x6	50	5	$269	1.3	$320	$416
2	6,832	8x8	160	10	283	1.3	333	433
3	10,248	10x8	240	15	297	1.3	348	453
4	13,664	10x10	320	20	315	1.4	370	481
5	17,080	12x10	400	24	324	1.4	379	492
7.5	25,620	16x10	600	36	363	1.5	421	547
10	34,160	18x12	800	48	505	1.7	571	742
15	51,240	30x12	1,200	72	551	2.0	629	817
20	68,320	26x18	1,600	96	691	2.3	781	1,015
25	85,400	40x18	2,000	120	851	2.8	960	1,248
208 OR 240 VOLT, 3 PHASE								
3	10,248	10x8	240	8	$240	1.3	$290	$378
5	17,080	12x10	400	12	333	1.4	388	504
10	34,160	18x12	800	24	434	1.7	500	650
15	51,240	30x12	1,200	36	511	2.0	589	765
20	68,320	26x18	1,600	48	638	2.3	728	946
30	102,480	40x18	2,400	72	810	2.6	911	1,184
50	170,800	48x24	4,000	120	1,119	2.8	1,228	1,597
75	256,200	60x30	6,000	180	1,591	3.8	1,739	2,261
100	341,600	60x36	8,000	240	2,065	5.0	2,260	2,937
480 VOLT, 3 PHASE								
5	17,080	12x10	400	6	$376	1.4	$431	$560
10	34,160	18x12	800	12	490	1.7	556	723
15	51,240	30x12	1,200	18	577	2.0	655	852
20	68,320	26x18	1,600	24	721	2.3	810	1,054
50	170,800	48x24	4,000	60	1,264	2.8	1,373	1,785
75	256,200	60x30	6,000	90	1,798	3.8	1,946	2,530
100	341,600	60x36	8,000	120	1,958	5.0	2,153	2,799
125	393,250	60x48	10,000	150	2,097	7.0	2,370	3,081
150	512,400	72x48	12,000	180	2,331	9.0	2,682	3,487

Direct labor costs are $39.00 per hour.

Estimating Duct Heaters
Gas Fired, Indoor
Includes Controls, Burners, Stainless Steel Exchanger, Electric Ignition

HEATING BtuH INPUT	OUTPUT	CFM 70 RISE	DIRECT MATERIAL COST		LABOR	TOTAL MATERIAL & LABOR	
			Each	Per 1,000 Btu	Man Hours	Direct Cost	With 30% O&P
25,000	20,000	325	$565	$22.61	4	$721	$938
50,000	40,000	650	651	13.02	5	846	1,100
75,000	60,000	975	710	9.47	5	905	1,177
100,000	80,000	1,300	770	7.70	6	1,004	1,305
150,000	120,000	2,000	1,061	7.07	7	1,334	1,734
200,000	160,000	2,600	1,217	6.08	8	1,529	1,987
250,000	200,000	3,300	1,575	6.30	9	1,926	2,503
300,000	240,000	4,000	1,772	5.91	10	2,162	2,810
350,000	280,000	4,600	1,928	5.51	11	2,357	3,065
400,000	320,000	5,300	2,099	5.25	12	2,567	3,337

Estimating Unit Heaters
Gas Fired, Suspended Indoors
Includes Burners, Gas Valve, Stainless Steel Heat Exchangers, Electric Ignition

HEATING BtuH INPUT	OUTPUT	CFM 70 RISE	DIRECT MATERIAL COST		LABOR	TOTAL MATERIAL & LABOR	
			Each	Per 1,000 Btu	Man Hours	Direct Cost	With 30% O&P
25,000	20,000	325	$525	$21.02	5	$720	$937
50,000	40,000	650	602	12.05	6	836	1,087
75,000	60,000	975	614	8.19	6	848	1,103
100,000	80,000	1,300	724	7.24	7	997	1,296
150,000	120,000	2,000	928	6.19	8	1,240	1,612
200,000	160,000	2,600	1,120	5.60	9	1,471	1,913
250,000	200,000	3,300	1,382	5.53	10	1,772	2,304
300,000	240,000	4,000	1,634	5.45	11	2,063	2,682
350,000	280,000	4,600	1,931	5.52	12	2,399	3,119
400,000	320,000	5,300	2,119	5.30	13	2,626	3,414

Add for flue, installed	$153.54
Add for thermostat & switch	$72.73
Add for typical electrical hookup	$80.81
Add for typical gas piping, 30 ft.	$323.23
Deduct for aluminum heat exchanger	10%

Direct labor costs are $39.00 per hour.

Gas Fired Cast Iron Boilers
Hot Water And Steam

BtuH (IBR RATINGS)			DIRECT MATERIAL COST		LABOR	TOTAL MATERIAL & LABOR	
Gross Output	Net Hot Water Output	Net Steam Output	Each	Per MBH	Man Hours	Direct Cost	With 30% O&P
146,000	127,000	110,000	$1,817	$12.45	16	$2,441	$3,174
251,000	218,000	188,000	2,586	10.30	23	3,483	4,527
352,000	306,000	264,000	3,394	9.64	26	4,408	5,730
450,000	391,000	337,000	3,919	8.71	29	5,050	6,565
548,000	477,000	411,000	5,251	9.58	31	6,460	8,398
830,000	722,000	622,000	7,635	9.20	37	9,078	11,802
1,012,000	880,000	759,000	8,081	7.98	46	9,875	12,837
1,394,000	1,212,000	1,056,000	11,152	8.00	50	13,102	17,032
1,624,000	1,412,000	1,248,000	13,091	8.06	54	15,197	19,756
1,980,000	1,722,000	1,537,000	15,676	7.92	65	18,211	23,675
2,320,000	2,017,000	1,801,000	18,425	7.94	74	21,311	27,704
2,784,000	2,421,000	2,162,000	21,657	7.78	82	24,855	32,311
3,092,000	2,689,000	2,400,000	23,757	7.68	88	27,189	35,346
3,710,000	3,226,000	2,880,000	27,960	7.54	95	31,665	41,165
4,330,000	3,765,000	3,361,000	32,809	7.58	112	37,177	48,330
4,950,000	4,304,000	3,843,000	36,525	7.38	122	41,283	53,668
6,180,000	5,374,000	4,800,000	54,949	8.89	144	60,565	78,735

1. Includes valves, controls, insulated jacket.
2. Oil fired hot water or steam versus gas fired, .96 multiplier on material.
3. Add for air eliminator package which includes expansion tank, air vent and fill valve.

110 - 264 MBH	$108.28
337 - 759 MBH	$119.60
1,056 - MBH & UP	$138.99

4. Gross output is the amount of heat needed to heat the building plus piping radiation losses in the distribution system plus pickup allowances for warm-ups, etc.
5. Net output is the amount needed to heat the building.

Direct labor costs are $39.00 per hour.

Baseboard Heating
Per Foot

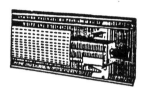

DESCRIPTION	SIZE	BtuH	MATERIAL COST	LABOR	TOTAL MATERIAL & LABOR PER FT	
			Per Ft	Man Hours	With Direct	30% O&P
HOT WATER						
RADIATION						
(Panel, Cast Iron, with Damper, Fin Tube, Wall Hung, Supports, Excluding Covers)			$25.53	0.8	$56.73	$73.75
ALUMINUM FIN						
Copper Tube	1-1/4		33.23	1	72.23	93.89
STEEL FIN						
Steel Tube	1-1/4		29.30	1	68.30	88.80
Steel Tube	2		33.30	1.3	84.00	109.20
PACKAGE						
ALUMINUM FIN						
Copper Tube	1/2		8.14	0.62	32.32	42.02
Copper Tube	3/4		8.51	0.67	34.64	45.03
Copper Tube	1		14.80	0.67	40.93	53.21
Copper Tube	1-1/4		21.09	0.67	47.22	61.39
STEEL FIN						
Iron Pipe Size						
Steel Tube			23.68	0.67	49.81	64.75
CONVECTOR UNIT						
(Damper, Flush, Trim, Floor Indented)			39.59	0.8	70.79	92.03

Infrared Units
Gas Fired, Electronic Ignition, No Vants, 100% Shut Off

MBH	DIRECT MATERIAL COST		LABOR	TOTAL MATERIAL & LABOR	
	Each	Per MBH	Man Hours	Direct Cost	With 30% O&P
15	$407	$27.13	2	$485	$631
30	451	15.05	3	568	739
45	548	12.17	3	665	864
50	562	11.25	4	718	934
60	629	10.48	4	785	1021
75	703	9.37	5	898	1167
90	784	8.72	6	1018	1324
105	925	8.81	8	1237	1608
120	977	8.14	8	1289	1675

NO PIPING OR WIRING INCLUDED

Electric Baseboard Heating
Commercial Grade, 187 W Per Ft, 641 Btu Per Ft

LENGTH FT	WATTS	BtuH	DIRECT MATERIAL COST		LABOR	TOTAL MATERIAL & LABOR		
			Each	Per Ft	Man Hours	Direct Costs	With 30% O&P	Per Ft
2	375	1,281	$37.00	18.50	1	$76.00	$98.80	$49.40
3	800	1,708	48.84	16.28	1	87.84	114.19	38.06
4	750	2,562	59.20	14.80	1.2	106.00	137.80	34.45
5	935	3,194	81.40	16.28	1.4	136.00	176.80	35.36
6	1,125	3,843	91.76	15.29	1.6	154.16		0.00
7	1,310	4,475	111.00	15.86	1.8	181.20	235.56	33.65
8	1,500	5,124	124.32	15.54	2	202.32	263.02	32.88
9	1,680	5,739	133.20	14.80	2.2	219.00	284.70	31.63
10	1,875	6,405	140.60	14.06	2.4	234.20	304.46	30.45

Electric Wall Heaters
Commercial

WATTS	BtuH	DIRECT MATERIAL COST Each	LABOR Man Hours	TOTAL MATERIAL & LABOR Direct Costs	With 30* O&P
750	2,562	$66.60	1.1	$109.50	$142.35
1,000	3,416	118.40	1.1	161.30	209.69
1,250	4,270	118.40	1.3	169.10	219.83
1,500	51124	177.60	1.6	240.00	312.00
2,000	6,832	185.00	1.6	247.40	321.62
2,500	8,540	192.40	2	270.40	351.52
3,000	10,248	199.80	2	277.80	361.14
4,000	13,664	214.60	2.3	304.30	395.59

Air Curtain Units
Steam And Hydronic Radiation Unit

AIR CURTAIN BLOWER UNITS MBH	LENGTH	CFM	DIRECT MATERIAL COST Each	Per CFM	LABOR Man Hours	TOTAL MATERIAL & LABOR Direct Cost	With 30% O&P
STEAM							
200	48" long,	3,980	$670.14	$0.17	16.2	$1,301	$1,691
230	60" long,	4,845	783.81	0.16	16.2	$1,414	$1,838
260	84" long,	3,050	1,017.06	0.33	20.0	$1,797	$2,336
285	96" long,	3,340	1,017.06	0.30	22.9	$1,911	$2,485
360	120" long,	4,320	1,207.68	0.28	28.5	$2,319	$3-.015
400	144" long,	5,000	1,361.60	0.27	34.0	$2,688	$3,494
530	120" long,	9,570	1,207.68	0.13	29.0	$2,339	$3,040
630	144" long,	11,375	1,361.60	0.12	35.5	$2,746	$3,570
HYDRONIC							
60	36" long,	1,370	670.14	0.49	6.5	$924	$1,201
75	48" long,	1,714	670.14	0.39	6.5	$924	$1,201
90	60" long,	2,055	742.37	0.36	6.5	$996	$1,295
100	361" long,	2,950	670.14	0.23	6.5	$924	$1,201
105	72" long,	2,550	833.54	0.33	8.1	$1,147	$1,492
130	84" long,	3,120	1,017.06	0.33	8.7	$1,356	$1,763
150	96" long,	3,420	688 ' 00	0.20	9.7	$1,066	$1,385
190	120" long,	4,200	1,207.68	0.29	12.9	$1,711	$2,224
230	144" long,	5,110	928.00	0.18	12.9	$1,431	$1,860
260	361" long,	7,700	1,018.24	0.13	9.7	$1,397	$1,816
300	120" long,	9,810	1,509.60	0.15	11.3	$1,950	$2,535
350	144" long,	11,375	1,702.00	0.15	12.9	$2,206	$2,868

NOTE
Costs include connecting units only and no electrical work.

DX Evaporator Coils
6 Row, 400 To 600 FPM
Air 75°F In, 55°F Out

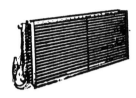

	BtuH	TONS	TYPICAL SIZE	SQ FT AREA 500 PPM	CFM @400 cfm Per Ton	DIRECT MATERIAL LABOR Each	Per Sq Ft	LABOR Man Hours	TOTAL MATERIAL Direct Costs	With 30% O&P		
	60,000	5	36x16	4	2,000	$567	$142	4	$723	$940		
	90,000	7.5	36x24	6	3,000	709	118	6	943	1,226	120,000	10
48x24	8	4,000	765	96	7	1,038	1,350	144,000	12	48x30	9.6	
4,800	895	93	8	1,207	1,570	180,000	15	48x36	12	6,000	1,051	88
10	1,441	1,873	240,000	20	48x48	16	8,000	1,159	72	12	1,627	
2,115	300,000	25	60x48	20	10,000	1,376	69	14	1,922	2,499		
	450,000	37.5	72x60	30	15,000	1,758	59	18	2,460	3,198	600,000	50
96x60	40	20,000	2,475	62	22	3,333	4,332	900,000	75	120x72	60	
30,000	3,175	53	27	4,228	5,496	1,200,000	100	120x96	80	40,000	3,777	47
32	5,025	6,532	1,500,000	125	120x120	100	50,000	4,558	46	37	6,001	
7,802	1,800,000	150	144x120	120	60,000	5,275	44	40	6,835	8,885	2,400,000	200
150x150	160	80,000	6,772	42	48	8,644	11,238					

CORRECTION FACTORS	MATERIAL	LABOR
1. 2 row coils	.75	.85
2. 4 row coils	.90	.92

Direct labor costs are $39.00 per hour.

Chilled Water Coils
6 Row With Drain Pan
400 To 600 FPS, Water 45°F In, 55°F Out

BtuH	TONS	TYPICAL SIZE	SQ FT AREA @ 500 fpm	GPM	CFM @400 cfm per ton	DIRECT MATERIAL COST Each	Per Sq Ft	LABOR Man Hours	TOTAL MATERIAL & LABOR Direct Costs	With 30% O&P
60,000	5	36x16	4	12	2,000	$515	$129	4	$671	$872
90 000	7.5	36x24	6	18	3,000	644	107	6	878	1,141
120,000	10	48x24	8	24	4,000	696	87	7	969	1,259
144,000	12	48x30	9.6	28.8	4,800	814	85	8	1,126	1,464
180,000	15	48x36	12	36	6,000	955	80	10	1,345	1,748
240,000	20	48x48	16	48	8,000	1,054	66	12	1,522	1,978
300,000	25	60x48	20	60	10,000	1,251	63	14	1,797	2,336
450,000	37.5	72x60	30	90	15,000	1,598	53	18	2,300	2,991
600,000	50	96x60	40	120	20,000	2,250	56	22	3,108	4,040
900,000	75	120x72	60	180	30,000	2,886	48	27	3,939	5,121
1,200,000	100	120x96	80	240	40,000	3,434	43	32	4,682	6,086
1,500,000	125	120x120	100	300	50,000	4,144	41	37	5,587	7,263
1,800,000	150	144x120	120	360	60,000	4,795	40	40	6,355	8,262
2,400,000	200	150x150	160	480	80,000	6,157	38	48	8,029	10,437

CORRECTION FACTORS	MATERIAL	LABOR
1. 2 row coils	.75	.85
2. 4 row coils	.90	.92

Direct labor costs are $39.00 per hour.

Centrifugal Water Cooled Chillers
With Condenser and Single Compressor

| TONS | DIRECT APPROX WEIGHT | LABOR MATERIAL COST | | TOTAL | MATERIAL & LABOR | |
		Each	Per Ton	Man Hours	Direct Cost	With 30% O&P
100	7,500	$68,687	$687	76	$71,651	$93,146
200	10,000	77,576	388	80	80,696	104,904
300	12,000	92,121	307	88	95,553	124,219
400	18,000	109,899	275	92	113,487	147,533
800	20,000	127,677	255	95	131,382	170,796
600	24,000	147,071	245	98	150,893	196,160
700	24,000	169,697	242	102	173,675	225,777
800	27,000	189,091	236	106	193,225	251,192
900	33,000	210,101	233	112	214,469	278,809
1,000	37,000	229,495	229	118	234,097	304,326
1,200	43,000	271,515	226	124	276,351	359,256
1,400		310,303	222	132	315,451	410,086
1,600		350,707	219	140	356,167	463,017

Labor includes receiving package chiller at job site, unloading, uncrating, setting in place, aligning, etc.
Labor does not include piping, valves or electrical hookup.
Crane rental costs not included.
Chillers with DOUBLE BUNDLE condensers run $115.00 more per ton.
Direct labor costs are $39.00 per hour.

Reciprocating Chillers
Water Cooled With Multiple Semi-hermetic Compressors

TONS	DIRECT MATERIAL COST		LABOR	TOTAL MATERIAL & LABOR	
	Each	Per Ton	Man Hours	Direct Cost	With 30% O&P
20	$15,192	$760	28	$16,284	$21,169
40	22,626	566	32	23,874	31,037
60	30,222	504	36	31,626	41,113
80	38,465	481	40	40,025	52,033
100	46,869	469	48	48,741	63,363
120	55,596	463	56	57,780	75,114
140	61,900	442	62	64,318	83,613
160	68,848	430	66	71,422	92,849
180	74,592	414	70	77,322	100,519

CORRECTION FACTORS   MATERIAL   LABOR

1.   Air cooled reciprocating chiller without condenser   0.72   0.75

Labor includes unloading from truck at job site, staging, uncrating, hoisting and setting into place, anchoring, aligning, starting and checkout.

Man hours do not include piping, valves or electrical hook.

Based on direct labor costs of $39.00.

Reciprocating Chillers
Air Cooled With Compressor

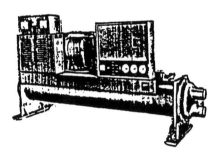

TONS	DIRECT MATERIAL COST		LABOR	TOTAL MATERIAL & LABOR	
	Each	Per Ton	Man Hours	Direct Cost	With 30% O&P
20	$16,970	$848	28	$18,062	$23,480
40	25,050	626	32	26,298	34,188
60	33,855	564	36	35,259	45,837
80	43,636	545	40	45,196	58,755
100	50,683	507	48	52,555	68,321
120	63,596	530	56	65,780	85,513

CORRECTION FACTORS   MATERIAL   LABOR

1. Air cooled reciprocating chiller without condenser   0.72   0.75

Labor includes unloading from truck at job site, staging, uncrating, hoisting andsetting into place, anchoring, aligning, starting and checkout.

Man hours do not include piping, valves or electrical hook.

Based on direct labor costs of $39.00.

Cooling Towers

TONS	GPM	FAN Hp	DIRECT MATERIAL COSTS		TOTAL LABOR	MATERIAL & LABOR	
			Each	Per Ton	Man Hours	Direct Costs	With 30% O&P
CROSSFLOW, INDUCED DRAFT, PROPELLER FAN, SHIPPED UNASSEMBLED*							
100	240	5	$5,658	$57	130	$10,728	$13,946
200	480	7.5	10,764	54	190	18,174	23,626
300	720	15	13,593	45	260	23,733	30,853
400	960	20	15,594	39	330	28,464	37,003
500	1,200	20	18,009	36	370	32,439	42,171
600	1,440	25	20,42 4	34	400	36,024	46,831
800	1,920	30	25,530	32	450	43,080	56,004
1,000	2,400	35	28,980	29	550	50,430	65,559
COUNTER FLOW, FORCED DRAFT, CENTRIFUGAL FAN, SHIPPED UNASSEMBLED							
100	240		$7,728	$77	137	$13,071	$16,992
200	480		14,766	74	200	22,566	29,336
300	720		17,250	58	273	27,897	36,266
400	960		22,770	57	347	36,303	47,194
500	1,200		27,738	55	389	42,909	55,782
600	1,440		32,706	55	420	49,086	63,612
800	1,920		43,470	54	473	61,917	80,492
1,000	2,400		54,510	55	578	77,052	100,168
INDUCED DRAFT, PROPELLER, GALVANIZED, SHIPPED ASSEMBLED							
10	24	750	$1,725	$173	12	$2,193	$2,851
15	36	900	2,484	166	13	2,991	3,888
20	48	980	3,174	159	14	3,720	4,836
25	60	1,000	3,795	152	16	4,419	5,745
30	72	1,180	4,347	145	17	5,010	6,513
40	96	1,290	5,520	138	18	6,222	8,089
50	120	1,800	6,072	121	20	6,852	8,908
60	144	2,300	6,293	105	22	7,151	9,296
80	192	3,500	7,066	88	24	8,002	10,402
100	240	4,300	8,280	83	27	9,333	12,133
125	300	4,900	10,005	80	29	11,136	14,477
150	360	6,000	11,592	77	32	12,840	16,692
175	420	6,200	13,041	75	34	14,367	18,677
200	480	6,900	14,352	72	36	15,756	20,483
300	720	9,000	20,700	69	51	22,689	29,496
400	960	14,500	26,496	66	63	28,953	37,639
800	1,200	16,900	31,740	63	72	34,548	44,912

*REDWOOD, TREATED FIR

Man hours include unloading, handling, assembling and set in place.
Piping, electrical wiring or crane rental are not included.
Direct labor costs are $39.00 per hour.

Heat Pumps

TONS OF COOLING	HEATING CAPACITY	DIRECT MATERIAL COST		LABOR	TOTAL MATERIAL & LABOR	
		Each	Per Ton	Man Hours	Direct Cost	With 30% O&P
SPLIT SYSTEM, AIR TO AIR						
2	8.5	$1,687	$570	9	$2,038	$2,650
3	13	2,324	523	10	2,714	3,528
5	27	3,904	528	20	4,684	6,090
7	33	6,323	610	24	7,259	9,436
10	50	8,295	561	26	9,309	12,102
15	64	11,458	516	30	12,628	16,417
25	119	18,567	502	40	20,127	26,165
30	163	26,714	602	44	28,430	36,959
40	193	29,533	499	50	31,483	40,928
PACKAGE UNIT, AIR TO AIR						
2	6.5	$1,616	$546	5	$1,811	$2,355
3	10	2,324	523	6	2,558	3,325
4	13	2,671	451	8	2,983	3,878
5	27	3,448	466	12	3,916	5,091
7	35	4,851	468	14	5,397	7,017
15	56	12,654	570	18	13,356	17,363
20	100	16,172	546	21	16,991	22,088
25	120	19,262	521	24	20,198	26,258
30	163	24,605	554	26	25,619	33,305
PACKAGE UNIT, WATER SOURCE TO AIR						
1	13	$1,191	$805	5	$1,386	$1,802
2	19	1,199	405	6	1,433	1,863
3	27	1,621	365	7	1,894	2,462
4	31	2,109	356	8	2,421	3,147
5	29	2,849	385	10	3,239	4,211
7.5	35	3,619	326	16	4,243	5,515
10	50	4,499	304	18	5,201	6,762
15	64	7,866	354	32	9,114	11,848
20	100	8,998	304	36	10,402	13,523

1. Add 5 percent on material for supplementary electric heating coils on air to air units.
2. For water source units add 10 percent to material.

Direct labor costs are $39.00 per hour.

Condensing Units
Air Cooled, Staged Compressors, Controls

TONS	CFM	DIRECT MATERIAL COST		LABOR	TOTAL MATERIAL & LABOR	
		Each	Per Ton	Man Hours	Direct Cost	With 30% O&P
3	1,200	$1,798	$599	4	$1,954	$2,540
5	2,000	3,083	617	6	3,317	4,312
7.5	3,000	4,008	534	8	4,320	5,616
10	4,000	5,344	534	9	5,695	7,404
12.5	5,000	6,271	502	10	6,661	8,659
15	6,000	7,194	480	12	7,662	9,961
20	8,000	9,558	478	14	10,104	13,136
25	10,000	11,906	476	14	12,452	16,187
30	12,000	14,184	473	is	14,866	19,351
40	16,000	18,844	471	22	19,702	25,613
50	20,000	23,554	471	24	24,490	31,837
60	24,000	28,163	469	30	29,333	36,133
80	32,000	37,550	469	36	38,954	50,641
100	40,000	46,939	469	42	48,577	63,151

This page intentionally left blank

Chapter 6

HVAC Units and
Air Distribution Equipment

The equipment tables in section III of this manual are designed to maximize the accuracy of the selections from the tables.

- The left columns in the tables show the size or capacity of the particular piece of equipment, alternate units of measurements if required and sometimes secondary factors-which altogether make selections more specific and realistic.

  Column one might be the capacity of the equipment in typical units such as GPM, BtuK KW's, watts, MBH, tons of cooling, CFM of air, etc. The size might also be denoted in actual dimensional units such as 24″ x 24″.

  The second factor in the sizing and capacity columns might involve a condition such as feet of head along with gpm as with pumps.

  In some cases, there may be another condition listed long with gpm and feet of head; such as the horsepower of a motor for a particular unit.

- The middle columns involve the direct material costs for each size plus a unit cost based on the units of measurement being used for the size category. For example, a certain GPM pump might have a unit material cost of $41.00 per gallon.

- The right columns show the labor in man hours, the total direct costs of labor and material together and the installed selling price with overhead and profit included.

Estimating Roof Top Units
Single Zone, DX Cooling, Electric Heating Coils
Includes Economizers, Coils, Filters, Curbs,
Standard Controls, Warranty

TONS	CFM	ELECTRIC COIL		DIRECT MATERIAL COST		LABOR	TOTAL MATERIAL & LABOR	
		Kw	Each	Per Ton	Per CFM	Man Hours	Direct Cost	With 30% O&P
2	800	15	$2,050	$1,025	$2.56	4	$2,206	$2,868
3	1,200	20	2,427	809	2.02	4	2,583	3,358
5	2,000	25	3,854	771	1.93	6	4,088	5,315
7.5	3,000	37	6,869	916	2.29	8	7,181	9,336
10	4,000	50	9,020	902	2.25	8	9,332	12,131
12.5	5,000	62	$11,178	$894	$2.24	10	$11,568	$15,038
15	6,000	75	13,206	880	2.20	10	13,596	17,675
20	8,000	87	17,268	863	2.16	12	17,736	23,057
25	10,000	100	21,199	848	2.12	12	21,667	28,168
30	12,000	- 125	$24,980	$833	$2.08	16	$25,604	$33,285
40	16,000	150	32,686	817	2.04	20	33,466	43,506
50	20,000	190	40,087	802	2.00	22	40,945	53,228
60	24,000	220	46,254	771	1.93	26	47,268	61,449

CORRECTION FACTORS	Material	Labor
1. Variable air volume unit	1.28	1.10
2. Multizone units	1.33	1.25
3. Heat Pumps (with electric heating coils)	1.30	--
4. Cooling only, no heating	0.66	--
5. Power return fan section, add direct costs		
10 tons	$1,747	1.10
20 tons	2,618	1.10
50 tons	5,762	1.10
6. Omit economizer, deduct		
10 tons	$1,134	0.95
20 tons	1,397	0.95
50 tons	2,095	0.95

Estimating Roof Top Units
Single Zone, DX Cooling, Gas Heating, Staged Cooling and Heating, 7-1/2 Ton and Up Includes Economizers, Coils, Filters, Curbs, Standard Controls, Warranty

TONS	CFM	HEATING MBH	Each	DIRECT MATERIAL COST Per Ton	DIRECT MATERIAL COST Per CFM	LABOR Man Hours	TOTAL MATERIAL & LABOR Direct Cost	TOTAL MATERIAL & LABOR With 30% O&P
2	800	60	$2,343	$1,172	$2.93	6	$2,577	$3,350
3	1,200	94	2,698	899	2.25	8	3,010	3,913
5	2,000	112	4,282	856	2.14	10	4,672	6,074
7.5	3,000	135	7,644	1,019	2.55	12	8,112	10,545
10	4,000	200	10,022	1,002	2.51	12	10,490	13,637
12.5	5,000	225	$12,420	$994	$2.48	14	$12,966	$16,856
15	6,000	270	14,672	978	2.45	14	15,218	19,783
20	8,000	360	19,186	959	2.40	16	19,810	25,753
25	10,000	450	23,551	942	2.36	16	24,175	31,428
30	12,000	540	$27,752	$925	$2.31	18	$28,454	$36,990
40	16,000	675	36,317	908	2.27	24	37,253	48,429
50	20,000	810	44,541	891	2.23	26	45,555	59,222
60	24,000	985	52,421	874	2.18	32	53,669	69,769

CORRECTION FACTORS	Material	Labor
1. Variable air volume unit	1.25	1.10
2. Multizone units	1.30	1.25
3. Hot water heating coils	0.90	--
4. Steam heating coil	0.93	--
5. Power return fan section, add direct costs		
10 tons	$1,647	1.10
20 tons	2,471	1.10
50 tons	5,436	1.10
6. Omit economizer, deduct		
10 tons	$1,071	0.95
20 tons	1,318	0.95
50 tons	1,977	0.95

Air Handling Units
DX Coll, Electric Heating Coll, Single Zone, Fan and Coll Section
Isolators, Throwaway Filters, Motors, Drives

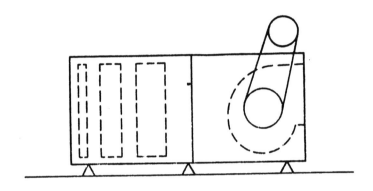

TONS	CFM	HEATING MBH	Each	DIRECT MATERIAL COST Per Ton	DIRECT MATERIAL COST Per CFM	LABOR Man Hours	TOTAL MATERIAL & LABOR Direct Cost	TOTAL MATERIAL & LABOR With 30% O&P
3	1,200	95	$1,998	$666	$1.66	4	$2,154	$2,800
5	2,000	112	3,118	624	1.56	9	3,469	4,510
7.5	3,000	135	4,387	585	1.46	10	4,777	6,210
10	4,000	200	5,581	558	1.40	12	6,049	7,864
12.5	5,000	225	$6,568	$525	$1.31	14	$7,114	$9,248
15	6,000	270	7,327	488	1.22	16	7,951	10,336
20	8,000	360	8,703	435	1.09	18	9,405	12,226
25	10,000	450	9,481	379	0.95	20	10,261	13,340
30	12,000	540	11,082	369	0.92	22	11,940	15,522
40	16,000	675	$13,545	$339	$0.85	26	$14,559	$18,927
50	20,000	810	15,803	316	0.79	30	16,973	22,065
60	24,000	984	18,472	308	0.77	34	19,798	25,737
80	32,000	1,312	23,644	296	0.74	42	25,282	32,866
100	40,000	1,640	28,324	283	0.71	50	30,274	39,357

Air conditioning only, no electric heating coil.
Direct material multiplier .75

Air Handling Units
DX Coil, Gas Burners,
Single Zone, Fan and Coil Section
Isolators, Throwaway Filters, Motors, Drives

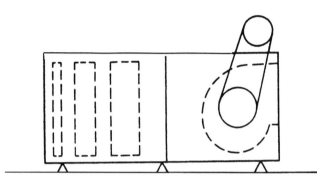

TONS	CFM	HEATING MBH	Each	DIRECT MATERIAL COST Per Ton	DIRECT MATERIAL COST Per CFM	LABOR Man Hours	TOTAL MATERIAL & LABOR Direct Cost	TOTAL MATERIAL & LABOR With 30% O&P
3	1,200	95	$2,503	$834	$2.09	4	$2,659	$3,456
5	2,000	112	3,905	781	1.95	9	4,256	5,533
7.5	3,000	135	5,492	732	1.83	10	5,882	7,646
10	4,000	200	6,990	699	1.75	12	7,458	9,695
12.5	5,000	225	$8,222	$658	$1.64	14	$8,768	$11,399
15	6,000	270	9,172	611	1.53	16	9,796	12,735
20	8,000	360	10,896	545	1.36	18	11,598	15,077
25	10,000	450	11,872	475	1.19	20	12,652	16,447
30	12,000	540	13,876	463	1.16	22	14,734	19,154
40	16,000	675	$16,959	$424	$1.06	26	$17,973	$23,365
50	20,000	810	19,786	396	0.99	30	20,956	27,243
60	24,000	984	23,126	385	0.96	34	24,452	31,788
80	32,000	1,312	29,602	370	0.93	42	31,240	40,612
100	40,000	1,640	35,462	355	0.89	50	37,412	48,635

Estimating Air Handling Units
2 Row Water Or Steam Heating Coil, 6 Row Chilled Water Coil
Single Zone, Fan and Coil Section
Isolators, Throwaway Filters, Motors, Drives

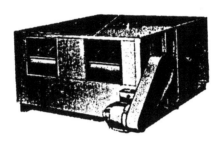

TONS	CFM	HEATING MBH	Each	DIRECT MATERIAL COST		LABOR	TOTAL MATERIAL & LABOR	
				Per Ton	Per CFM	Man Hours	Direct Cost	With 30% O&P
3	1,200	95	$1,667	$556	$1.39	4	$1,823	$2,370
5	2,000	112	2,603	521	1.30	9	2,954	3,840
7.5	3,000	135	3,662	488	1.22	10	4,052	5,267
10	4,000	200	4,659	466	1.16	12	5,127	6,665
12.5	5,000	225	$5,481	$438	$1.10	14	$6,027	$7,835
15	6,000	270	6,115	408	1.02	16	6,739	8,761
20	8,000	360	7,262	363	0.91	18	7,964	10,353
25	10,000	450	7,913	317	0.79	20	8,693	11,301
30	12,000	540	9,250	308	0.77	22	10,108	13,140
40	16,000	675	$11,305	$283	$0.71	26	$12,319	$16,015
50	20,000	810	13,191	264	0.66	30	14,361	18,669
60	24,000	984	15,418	257	0.64	34	16,744	21,767
80	32,000	1,312	19,734	247	0.62	42	21,372	27,784
100	40,000	1,640	23,641	236	0.59	50	25,591	33,268

CORRECTION FACTORS ON ALL AIR HANDLING UNITS		Material		Labor
1. Multi-zone unit instead of single zone,				
	DX and Electric	1.40		1.25
	DX and Gas	1.32		1.25
	HW and CHW Coils	1.50		1.25
2. Suspended installation instead of floor mounted		1.25		
3. With filter, mixing box (add)		0.31	/CFM	1.05
4. Variable Air Volume (add)		0.55	/CFM	1.10
5. With economizer section (add)	10 tons	$1,071		1.05
	20 tons	1,318		1.05
	50 tons	1,977		1.05

Self Contained Air Conditioning Units
Air Cooled, DX Coil, Condenser Section, Electric Heating Coil, Supply Fan, Filters

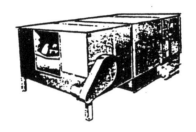

TONS	CFM	HEATING MBH	Each	DIRECT MATERIAL COST Per Ton	DIRECT MATERIAL COST Per CFM	LABOR Man Hours	TOTAL MATERIAL & LABOR Direct Cost	TOTAL MATERIAL & LABOR With 30% O&P
3	1,200	94	$3,212	$1,071	$2.68	6	$3,446	$4,480
5	2,000	115	3,691	738	1.85	12	4,159	5,407
7.5	3,000	135	5,292	706	1.76	14	5,838	7,589
10	4,000	200	6,783	678	1.70	15	7,368	9,579
12.5	5,000	225	8,352	668	1.67	18	9,054	11,770
15	6,000	270	9,868	658	1.64	20	10,648	13,842
20	8,000	360	12,915	646	1.61	22	13,773	17,905
25	10,000	450	15,204	608	1.52	24	16,140	20,982
30	12,000	540	17,473	582	1.46	26	18,487	24,033
40	16,000	675	21,584	540	1.35	31	22,793	29,631
50	20,000	810	25,097	502	1.25	36	26,501	34,451
60	24,000	985	29,808	497	1.24	42	31,446	40,880

CORRECTION FACTORS   Material   Labor
1.   Water cooled
2.   Water or steam heating coils.
Direct labor costs are $39.00 per hour.

Room Heating and Cooling
Unit Ventilators
2 Pipe, Single Coil, Controls, Filters, Floor Mounted

BtuH COOLING	DIRECT MATERIAL COST		TOTAL LABOR	MATERIAL & LABOR	
	Per Each	Btu	Man Hours	Direct Cost	With 30% O&P
6,000	$1,924	$0.321	5.0	$2,119	$2,755
9,000	1,961	0.218	5.5	2,176	2,828
12,000	2,109	0.176	6.1	2,347	3,051
18,000	2,294	0.127	7.0	2,567	3,337
24,000	2,620	0.109	9.3	2,982	3,877
30,000	2,997	0.100	11.5	3,446	4,479
36,000	3,071	0.085	13.2	3,586	4,662
42,000	3,182	0.076	18.0	3,884	5,049
48,000	3,293	0.069	23.2	4,198	5,457
60,000	3,959	0.066	37.0	5,402	7,023

CORRECTION FACTORS 1. 4 pipe, 2 coil, heating and cooling, +14%. 2. For separate electric heating coil, +30%. 3. Ceiling hung unit ventilators, +5%.

Estimating Centrifugal Fans
Air Foil Wheels, Single Width Single Inlet
Outlet Velocity 2000 FPM
Includes Motors, Drives, Isolators

TONS	CFM	WHEEL DIAMETER		DIRECT MATERIAL COST		LABOR	TOTAL MATERIAL & LABOR	
			Each	Per Ton	Per CFM	Man Hours	Direct Cost	With 30% O&P
1,000	1/3	12	1″	$753	$0.75	3	$870	$1,131
2,000	1/2	15	1″	1,358	0.68	7	1,631	2,120
4,000	2	18	1″	1,807	0.45	10	2,197	2,856
6,000	2-1/2	22	1-1/2″	2,260	0.38	12	2,728	3,546
8,000	3	27	1-1/2′	2,863	0.36	14	3,409	4,432
10,000	5	30	2″	3,391	0.34	16	4,015	5,219
12,000	7-1/2	33	2″	3,988	0.33	18	4,690	6,097
14,000	7-1/2	33	2″	4,747	0.34	20	5,527	7,185
16,000	7-1/2	36-1/2	2″	5,126	0.32	22	5,984	7,779
18,000	10	40-1/4	2″	5,766	0.32	24	6,702	8,712
20,000	10	44-1/4	2″	6,029	0.30	26	7,043	9,156
25,000	15	44-1/4	2-1/2″	7,537	0.30	30	8,707	11,320
30,000	20	54-1/4	2-1/2″	9,044	0.30	36	10,448	13,583
40,000	25	60	2-1/2″	11,305	0.28	46	13,099	17,029
50,000	30	66	3″	14,133	0.28	56	16,317	21,212
60,000	40	73	3″	16,959	0.28	64	19,455	25,291

Estimating Utility Sets
Forward Curve Wheels, Single Width Single Inlet
Includes Motors, Drives, Isolators

TONS	CFM	WHEEL DIAMETER		DIRECT MATERIAL COST		LABOR	TOTAL MATERIAL & LABOR	
			Each	Per Ton	Per CFM	Man Hours	Direct Cost	With 30% O&P
500	1/4	10	1/2″	$376	$0.75	2	$454	$590
1,000	1/2	12	3/4″	639	0.64	4	795	1,033
2,000	1	12	1″	1,054	0.53	5	1,249	1,624
4,000	2	18	1″	1,205	0.30	7	1,478	1,922
6,000	3	24	1-1/4″	1,695	0.28	9	2,046	2,659
8,000	5	27	1-1/2″	1,959	0.24	10	2,349	3,054
10,000	5	30	1-1/2″	2,260	0.23	12	2,728	3,546
12,000	5	33	1-1/2″	2,712	0.23	14	3,258	4,235
14,000	7-1/2	36	1-1/2″	3,165	0.23	16	3,789	4,925
16,000	7-1/2	39	1-1/2″	3,315	0.21	18	4,017	5,223
18,000	10	40	1-1/2″	3,731	0.21	20	4,511	5,864
20,000	15	44	1-1/2″	4,145	0.21	22	5,003	6,504

Estimating Roof Exhaust Fans
Centrifugal, Belt Driven, Aluminum Housing
1/2″ Static Pressure with Shutter, Birdscreen, Curb

TONS	CFM	WHEEL DIAMETER	DIRECT MATERIAL COST		LABOR	TOTAL MATERIAL & LABOR	
			Each	Per CFM	Man Hours	Direct Cost	With 30% O&P
500	1/12	10	$665	$1.33	3	$782	$1,017
1,000	1/6	12	793	0.79	3	910	1,183
1,500	1/4	14	797	0.53	3	914	1,189
2,000	1/3	22	980	0.49	4	1,136	1,477
3,000	1/2	24	1,110	0.37	4	1,266	1,645
4,000	3/4	24	1,281	0.32	5	1,476	1,918
6,000	1	30	1,525	0.25	6	1,759	2,287
8,000	1-1/2	30	1,807	0.23	6	2,041	2,653
10,000	2	36	2,203	0.22	7	2,476	3,218
15,000	3	48	3,308	0.22	8	3,620	4,706
20,000	5	48	4,554	0.23	9	4,905	6,377

Estimating Vane Axial Fans
Automatic Controllable Pitch, Direct Drive
Includes Inlet Cones, T-Frame Motors, Horizontal Supports, Isolators, Pneumatic Actuator

CFM	HP	FAN SIZE	INCHES STATIC PRESSURE	DIRECT MATERIAL COST		LABOR	TOTAL MATERIAL & LABOR	
				Each	Per CFM	Man Hours	Direct Cost	With 30% O&P
Supply Fans Medium Pressure, 1,770 RPM								
20,000	30	36/26	5	$13,490	$0.67	22	$14,348	$18,653
40,000	50	42/26	5	15,097	0.38	32	16,345	21,249
60,000	100	48/26	6	18,752	0.31	44	20,468	26,608
80,000	150	48/26	6	21,357	0.27	48	23,229	30,198
110,000	200	54/26	6	26,513	0.24	64	29,009	37,711
Return Air Fans, 1,170 RPM								
18,000	10	38/26	2	12,712	0.71	20	13,492	17,540
36,000	20	45/26	2	14,064	0.39	36.	15,468	20,108
54,000	30	54/26	2	15,815	0.29	48	17,687	22,993
72,000	50	54/26	2	17,262	0.24	50	19,212	24,976
100,000	75	60/26	2	22,132	0.22	64	24,628	32,017

Above fan selection and prices based on Joy vane-axial fans

Direct Labor costs are $39.00 per hour.

Estimating Multiblade Dampers
Automatic Control and Manual Multiblade Dampers, Opposed Blade with Frames and Bearings

SIZE	SEMI-PERIM	SQ FT	DIRECT MATERIAL COST		LABOR	TOTAL MATERIAL & LABOR	
	Inches		Each	Per Sq Ft	Man Hours	Direct Cost	With 30% O&P
12x6	18	0.5	$77.09	$154.19	1.0	$116	$151
12x12	24	1.0	80.51	80.51	1.1	123	160
18x12	30	1.5	89.07	59.38	1.3	140	182
18x18	36	2.3	98.51	42.83	1.3	149	194
24x12	36	2.0	90.80	45.40	1.4	145	189
24x24	48	4.0	133.61	33.40	1.8	204	265
30x12	42	2.5	$104.95	$41.98	1.5	$163	$212
30x18	48	3.8	130.20	34.26	1.8	200	261
30x24	54	5.0	147.33	29.47	2.0	225	293
36x18	54	4.5	141.10	31.36	2.0	219	285
36x36	72	9.0	153.56	17.06	2.4	247	321
42x18	60	6.0	153.56	25.59	2.2	239	311
42x24	66	7.0	167.88	23.98	2.4	261	340
48x24	72	8.0	185.01	23.13	2.6	286	372
48x48	96	16.0	354.62	22.16	3.5	491	638
54x18	72	6.8	$162.76	$23.93	2.6	$264	$343
54x24	78	9.0	207.29	23.03	2.8	316	411
60x24	84	10.0	227.85	22.78	3.0	345	448
60x48	108	20.0	438.57	21.93	4.1	598	778
72x24	96	12.0	268.96	22.41	3.5	405	527
72x48	120	24.0	524.22	21.84	4.3	692	899
72x72	144	36.0	709.10	19.70	5.5	924	1,201
84x36	120	21.0	$459.11	$21.86	4.3	$627	$815
84x84	168	49.0	1048.45	21.40	6.0	1,282	1,667
96x48	144	32.0	690.39	21.57	5.6	909	1,181
96x96	192	64.0	1370.51	21.41	6.0	1,605	2,086
120x60	180	50.0	1070.72	21.41	6.5	1,324	1,721
120x96	216	80.0	1,713.13	21.41	8.5	2,045	2,658

CORRECTION FACTORS	Material	Labor
1. Manual multiblade dampers	0.90	1.00

Estimating Fire Dampers
Curtain Type Blades, Vertical Installation

SIZE	SEMI-PERIM	SQ FT	DIRECT MATERIAL COST		LABOR	TOTAL MATERIAL & LABOR	
	Inches		Each	Per Sq Ft	Man Hours	Direct Cost	With 30% O&P
12x6	18	0.5	$25.93	$51.86	1.0	$65	$84
12x12	24	1.0	29.13	29.13	1.1	72	94
18x12	30	1.5	38.54	25.69	1.3	89	116
18x18	36	2.3	49.68	21.60	1.3	100	130
24x12	36	2.0	45.24	22.62	1.4	100	130
24x24	48	4.0	75.38	18.84	1.8	146	189
30x12	42	2.5	$52.26	$20.90	1.5	$111	$144
30x18	48	3.8	71.62	18.85	1.8	142	184
30x24	54	5.0	94.23	18.85	2.0	172	224
36x18	54	4.5	80.94	17.99	2.0	159	207
36x36	72	9.0	123.36	13.71	2.4	217	282
42x24	66	7.0	106.06	15.15	2.4	200	260
48x24	72	8.0	114.43	14.30	2.6	216	281
48x48	96	16.0	191.88	11.99	3.5	328	427
54x18	72	6.8	$103.10	$15.16	2.6	$204	$266
54x24	78	9.0	123.36	13.71	2.8	233	302
60x24	84	10.0	133.61	13.36	3.0	251	326
60x48	108	20.0	226.14	11.31	4.1	386	502

CORRECTION FACTORS	Material	Labor
1. U.L. Labels	1.10	---
2. Horizontal installation	1.20	---
3. 22 ga. U.L. sleeve	1.40	1.20
4. 22 ga. U.L. sleeve, free area	1.60	1.25
5. Cap for blades out of air stream	1.25	1.25

6. Multiblade type fire dampers, parallel
   blades Material factor of 1.1 times
   MULTIBLADE CONTROL DAMPER prices.

Estimating Louvers
Extruded Aluminum, 4" Deep, Fixed Blade,
Storm Proof with Screens, Mill Finish

SIZE	SEMI-PERIM	SQ FT	DIRECT MATERIAL COST		LABOR	TOTAL MATERIAL & LABOR	
	Inches		Each	Per Sq Ft	Man Hours	Direct Cost	With 30% O&P
12x6	18	0.5	$48.48	$96.97	1.1	$91	$119
12x12	24	1.0	56.57	56.57	1.2	103	134
18x12	30	1.5	78.80	52.53	1.3	129	168
18x18	36	2.3	107.80	46.87	1.5	166	216
24x12	36	2.0	96.97	48.48	1.5	155	202
24x18	42	3.0	130.91	43.64	2.0	209	272
30x12	42	2.5	$115.14	$46.06	2.0	$193	$251
30x18	48	3.8	147.39	38.79	2.1	229	298
30x24	54	5.0	169.70	33.94	2.3	259	337
36x18	54	4.5	163.63	36.36	2.3	253	329
36x36	72	9.0	255.27	28.36	2.6	357	464
42x18	60	6.0	193.94	32.32	2.4	288	374
42x24	66	7.0	217.20	31.03	2.6	319	414
48x24	72	8.0	237.26	29.66	2.8	346	450
48x48	96	16.0	404.69	25.29	4.0	561	729
54x18	72	6.8	$208.81	$30.71	2.8	$318	$413
54x24	78	10.0	269.89	26.99	3.3	399	518
60x24	84	10.0	269.89	26.99	3.5	406	528
60x48	108	20.0	488.07	24.40	4.5	664	863
72x24	96	12.0	310.30	25.86	4.0	466	606
72x48	120	24.0	566.31	23.60	5.0	761	990
72x72	144	36.0	843.63	23.43	6.0	1,078	1,401
84x36	120	21.0	$509.09	$24.24	5.0	$704	$915
84x84	168	49.0	1,148.29	23.43	7.0	1,421	1,848
96x48	144	32.0	749.90	23.43	6.0	984	1,279
96x96	192	64.0	1,499.80	23.43	8.0	1,812	2,355
120x60	180	50.0	1,171.72	23.43	7.5	1,464	1,903
120x96	216	80.0	1,874.75	23.43	9.0	2,226	2,893

Labor based on scaffold height installation in basement or on 1st floor or man height installation in penthouse wall.

CORRECTION FACTORS	Material	Labor
1. Galvanized		1.20
2. Finishes: baked enamel add	$3.51 / sq ft	
anodized, add	4.41 / sq ft	
duranodic, add	7.90 / sq ft	
fluoropolymer coating, add	12.17 / sq ft	

Estimating Registers
Return Air Registers, Fixed 450 Vanes, Opposed Blade Dampers, Commercial Grade

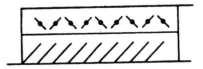

SIZE	SEMI-PERIM	SQ FT	DIRECT MATERIAL COST		LABOR	TOTAL MATERIAL & LABOR	
	Inches		Each	Per Sq Ft	Man Hours	Direct Cost	With 30% O&P
12x6	18	0.5	$30.83	$61.66	0.8	$62	$81
12x12	24	1.0	40.27	40.27	0.9	75	98
18x12	30	1.5	54.66	36.44	1.0	94	122
18x18	36	2.3	74.89	32.56	1.1	118	153
24x12	36	2.0	65.09	32.55	1.1	108	140
24x24	48	4.0	123.36	30.84	1.3	174	226
30x12	42	2.5	$82.44	$32.97	1.2	$129	$168
30x18	48	3.8	121.09	31.87	1.3	172	223
30x24	54	5.0	149.91	29.98	1.5	208	271
36x18	54	4.5	136.90	30.42	1.5	195	254
36x36	72	9.0	250.28	27.81	1.7	317	412
42x24	66	7.0	202.67	28.95	1.6	265	345
48x24	72	8.0	226.14	28.27	1.7	292	380
48x48	96	16.0	484.82	30.30	2.6	586	762
54x18	72	6.8	$197.02	$28.97	1.7	$263	$342
54x24	78	9.0	251.33	27.93	2.0	329	428
60x24	84	10.0	290.61	29.06	2.3	380	494
60x48	108	20.0	531.07	26.55	3.0	648	842

CORRECTION FACTORS—Commercial Grades	Material	Labor
1. Supply registers, single deflection dampers	1.10	1.00
2. Supply registers, double deflection dampers	1.25	1.00
3. Transfer grille, single deflection, no dampers........	0.63	0.60
4. Relief grille in ceiling, lay in, single deflection, no dampers	0.63	0.40
5. Aluminum construction instead of steel	1.10	1.00
6. Lay in type grilles...	1.00	0.50
7. Side wall screw in	1.00	0.65
8. Residential light commercial grade.........................	0.60	0.95

Ceiling Diffusers
Round, Fixed Pattern, Steel,
With Opposed Blade Dampers, Commercial Grade

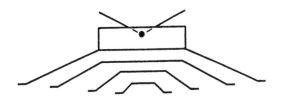

NECK DIAMETER	DIRECT MATERIAL COST		LABOR	TOTAL MATERIAL & LABOR	
Inches	Per In. Each	Man of Dia.	Direct Hours	With Cost	30% O&P
6	$46.87	$7.81	0.7	$74.17	$96.42
8	59.79	7.47	0.8	90.99	118.29
10	71.11	7.11	0.9	106.21	138.08
12	82.42	6.87	1.0	121.42	157.85
14	109.90	7.85	1.1	152.80	198.65
16	140.60	8.79	1.2	187.40	243.62
18	171.31	9.52	1.3	222.01	288.61
20	197.17	9.86	1.5	255.67	332.37
24	236.61	9.86	1.6	299.01	388.71
30	295.76	9.86	1.9	369.86	480.82
36	354.90	9.86	2.2	440.70	572.92

Round, Adjustable Pattern, Steel, W/OBD, Commercial Grade

6	$54.95	$9.16	0.7	$82.25	$106.93
8	69.50	8.69	0.8	100.70	130.91
10	82.42	8.24	0.9	117.52	152.78
12	95.36	7.95	1.0	134.36	174.66
14	127.68	9.12	1.1	170.58	221.75
16	163.23	10.20	1.2	210.03	273.04
15	198.79	11.04	1.3	249.49	324.34
20	229.49	11.47	1.5	287.99	374.39
24	275.40	11.47	1.6	337.80	439.14
30	344.25	11.47	1.9	418.35	543.85
36	413.73	11.49	2.2	499.53	649.39

Rectangular, Adjustable Pattern, Steel, w/OBD, Commercial G

6x6	$34.26	----	0.7	$61.56	$80.03
9x9	41.69	----	0.9	76.79	99.83
12x12	52.04	----	1.1	94.94	123.42
15x15	44.13	----	1.2	90.93	118.21
18x18	110.22	----	1.5	168.72	219.33
21x21	133.98	----	1.6	196.38	255.30
24x24	155.96	----	1.7	222.26	288.94

CORRECTION FACTORS	MATERIAL	LABOR
1. Aluminum	1.10	1.00
2. Layin diffusers	0.65	0.60

VAV Terminal Boxes
Cooling Only with Pneumatic or Electric Motor, Controls

CFM RANGE	COIL SQ FT	DIRECT MATERIAL COST	LABOR Man Hours	DIRECT MATERIAL & LABOR Direct Cost	With 30% O&P
200 - 400	----	$335.55	2.0	$414	$538
400 - 600	----	395.57	2.7	501	651
600 - 800	----	460.22	3.3	589	766
800 - 1,000	----	511.01	4.0	667	867
1,000 - 1,500	----	569.50	4.5	745	969
1,500 - 2,000	----	638.77	5.0	834	1,084
2,000 - 3,000	----	692.64	5.5	907	1,179

VAV Terminal Boxes
With Reheat Coils, Pneumatic or Electric Motor, Controls

CFM RANGE	COIL SQ FT	DIRECT MATERIAL COST	LABOR Man Hours	DIRECT MATERIAL & LABOR Direct Cost	With 30% O&P
200 - 400	0.8	$484.85	2.7	$590	$767
400 - 600	1.0	557.19	2.9	670	871
600 - 800	1.5	671.09	3.6	811	1,055
800 - 1,000	2.0	737.28	4.3	905	1,176
1,000 - 1,500	2.5	912.75	4.9	1,104	1,435
1,500 - 2,000	3.0	1005.10	5.5	1,220	1,585
2,000 - 3,000	3.5	1089.75	6.0	1,324	1,721

VAV System Components

TITUS VAV BOX RETROFIT KITS   Cost each

1.   For Buensod or Tutle and Baily CAV boxes

                        up to 1,000 CFM $370

2.   Titus kits for Titus boxes

                        0 - 500 CFM $266
                        500 - 1,000 CFM $370

CARRIER VAV UNITS
1.   Moduline cooling only VAV boxes

                        with 3 slot diffusers...$252

VAV CONTROL PANELS
1.   Controls and monitors supply fans, return fans,
      static pressure sensor, supply and return air, air
      monitoring stations, transmitters, outside air control ....$8,880 to $14,800

2.   Static pressure regulator $222

VAV Ceiling Diffusers

DIA.	CFM	DIRECT COST	LABOR	DIRECT MATERIAL & LABOR	
Inches	Range	Each	Man Hours	Direct Cost	With 30% O&P
6	100-220	$252	2	$330	$428
8	160-355	252	2	330	428
10	260-580	252	2	330	428
12	380-890	252	2	330	428

Self contained variable volume ceiling diffusers with thermal sensors for discharge air and warm-up, built in volume controls and built in pressure sensor for automatic switch-over between cooling and heating.

Manufacturers
1. Therma-Fuser by Acutherm, Novato, CA.

Direct labor costs are $39.00 per hour.

Inlet Vane Dampers
For Centrifugal Fans

DIA.	AREA	DIRECT MATERIAL COST		LABOR	TOTAL MATERIAL & LABOR	
Inches	Sq ft	Each	Per Sq Ft	Man Hours	Direct Cost	With 30% O&P
18	1.8	$104	$57.72	1.5	$162	$211
20	2.2	114	51.80	1.7	180	234
22	2.6	126	48.54	1.9	200	260
24	3.1	140	45.14	2.0	218	283
26	3.7	153	41.44	2.2	239	311
28	4.3	173	40.33	2.3	263	342
30	5.0	190	38.07	2.4	284	369
33	6.0	214	35.66	2.6	315	410
36	7.1	242	34.04	2.8	351	456
39	8.3	270	32.56	3.1	391	508
42	9.6	298	31.08	3.3	427	555
45	11.0	326	29.60	3.5	462	601
48	12.6	354	28.12	4.0	510	663
54	16.0	424	26.47	4.6	603	784
60	19.6	516	26.33	5.3	723	940
66	23.8	609	25.60	6.0	843	1,096
72	28.2	712	25.25	6.5	966	1,255
78	33.2	831	25.04	7.1	1,108	1,441
84	38.5	963	25.01	7.6	1,259	1,637
90	44.1	1,099	24.92	7.8	1,403	1,824
96	50.3	1,243	24.72	8.0	1,555	2,022

Direct labor costs are $39.00 per hour.

Fixed Pitch Sheaves

PITCH DIA. B BELTS	DIRECT MATERIAL COST		LABOR	TOTAL MATERIAL & LABOR	
	Pulley	Bushing	Man Hours	Direct Cost	With 30% O&P
FIXED, SINGLE GROOVE, NO BUSHING					
1.9	$8.88	----	0.2	$16.68	$21.68
2.3	9.21	----	0.2	17.01	22.11
2.6	9.95	----	0.3	21.65	28.14
3.0	10.82	----	0.3	22.52	29.27
3.2	13.59	----	0.4	29.19	37.94
3.4	14.65	----	0.4	30.25	39.33
3.9	15.47	----	0.4	31.07	40.39
4.6	17.23	----	0.5	36.73	47.75
FIXED, DOUBLE GROOVE, BUSHINGS					
3.4	$27.08	15.688	0.4	$58.37	$75.88
3.8	29.48	15.688	0.4	60.77	79.00
4.6	36.72	15.688	0.4	68.01	88.41
4.8	37.70	15.688	0.4	68.98	89.68
5.4	41.53	15.688	0.4	72.82	94.66
5.8	44.09	15.688	0.5	79.28	103.06
6.4	47.89	15.688	0.5	83.08	108.01
6.8	49.20	15.688	0.5	84.38	109.70
7.0	49.49	15.688	0.5	84.68	110.08
8.0	50.72	15.688	0.6	89.81	116.75
9.0	52.61	15.688	0.6	91.70	119.21
11.0	62.34	15.688	0.6	101.43	131.85
12.4	67.33	15.688	0.7	110.31	143.41
13.6	80.14	15.688	0.7	123.13	160.07
15.4	110.59	15.688	0.7	153.57	199.65
18.4	146.43	15.688	0.7	189.42	246.24
FIXED, TRIPLE GROOVE, BUSHINGS					
3.4	$30.01	15.688	0.4	$61.30	$79.69
3.8	33.36	15.688	0.4	64.65	84.04
4.0	34.56	15.688	0.4	65.85	85.60
4.4	37.12	15.688	0.4	68.41	88.93
4.8	44.09	15.688	0.4	75.38	97.99
5.0	45.30	15.688	0.5	80.49	104.64
5.4	39.77	15.688	0.5	74.96	97.44
5.8	42.46	15.688	0.5	77.65	100.94
6.0	43.72	15.688	0.5	78.91	102.58
6.4	47.43	15.688	0.6	86.52	112.48
6.8	49.91	15.688	0.6	88.99	115.69
7.0	55.87	15.688	0.6	94.96	123.45
8.0	59.61	15,688	0.6	98.70	128.31
9.0	67.13	15.688	0.6	106.22	138.09
11.0	82.67	15.688	0.7	125.66	163.36
12.4	95.52	15.688	0.7	138.51	180.06
13.6	102.89	15.688	0.7	145.88	189.64
15.4	131.26	15.688	0.7	174.25	226.52
18.4	158.98	15.688	0.7	201.97	262.56

Direct labor costs are $39.00 per-hour.

Variable Pitch Sheaves

PITCH DIA. B BELTS	DIRECT COST	LABOR	TOTAL MATERIAL & LABOR	
		Man Hours	Direct Cost	With 30% O&P
VARIABLE PITCH SINGLE GROOVE				
2.4-3.2	$9.06	0.4	$24.66	$32.05
2.7-3.7	11.96	0.4	27.56	35.83
3.1-4.1	14.13	0.4	29.73	38.65
3.7-4.7	32.19	0.5	51.69	67.20
4.3-5.3	43.51	0.5	63.01	81.92
4.9-5.9	53.12	0.6	76.52	99.47
5.5-6.5	55.26	0.6	78.66	102.26
VARIABLE PITCH, DOUBLE GROOVE				
2.5-3.3	$42.11	0.5	$61.61	$80.09
2.9-3.9	48.19	0.5	67.69	88.00
3.7-4.7	55.50	0.6	78.90	102.57
4.3-5.3	65.42	0.6	88.82	115.46
4.9-5.9	83.49	0.7	110.79	144.02
5.5-6.5	89.47	0.7	116.77	151.80

Direct labor costs are $39.00 per hour.

V-Belts

DIA.	DIRECT COST	LABOR	TOTAL MATERIAL & LABOR	
	Each	Man Hours	Direct Cost	With 30% O&P
33	$6.62	0.1	$10.52	$13.67
38	7.24	0.1	11.14	14.48
43	8.23	0.1	12.13	15.77
47	9.06	0.1	12.96	16.84
53	9.99	0.2	17.79	23.13
57	10.43	0.2	18.23	23.70
63	10.83	0.2	18.63	24.22
67	11.54	0.2	19.34	25.15
73	12.24	0.3	23.94	31.12
76	12.54	0.3	24.24	31.51
83	13.94	0.3	25.64	33.33
86	14.58	0.3	26.28	34.16
93	15.47	0.4	31.07	40.39
96	16.16	0.4	31.76	41.29
103	17.27	0.4	32.87	42.73
106	17.69	0.4	33.29	43.27
108	18.12	0.5	37.62	48.90
111	18.72	0.5	38.22	49.69
115	19.58	0.5	39.08	50.80
123	20.54	0.5	40.04	52.06
131	21.90	0.6	45.30	58.90
139	23.53	0.6	46.93	61.01
147	24.91	0.6	48.31	62.80
161	27.07	0.6	50.47	65.61
176	29.54	0.6	52.94	68.82

Direct labor costs are $39.00 per hour.

Filter Labor

	HOURS EACH
THROW AWAY AND PERMANENT FILTERS	
Slid Into Channels, I" or 2" thick	0.1
Assembled Into Filter Bank	0.5
GREASE FILTERS	0.2
(Laid in place in kitchen hood)	
BAG TYPE CARTRIDGES WITH FRAME	
(Shipped Unassembled)	
18x18	2.0
24x24	2.0
30x30	2.5
HEPA. FILTER CARTRIDGES	
24x12x6 deep	2.0
24x12x12 deep	2.4
24x24x6 deep	2.4
24x24x12 deep	2.8
CHARCOAL FILTER CARTRIDGES	
12x12	1.5
24x12	2.0
24x24	2.5
ROLL TYPE FILTER UNITS	
(Shipped Unassembled)	
One section up to 5 ft. wide	12.0
Two section filter bank	21.0
ELECTRO-STATIC FILTER UNITS	
24x12	2.0
24x24	2.5
36x24	4.0
48x36	8.0
60x48	12.0
72x60	14.0
84x72	16.0

Chapter 7

Plumbing Fixtures and Specialties

This chapter contains labor tables for plumbing fixtures such as:

*   Lavatories and Sinks
*   Toilets and Urinals
*   Drinking Fountains
*   Tubs and Showers
*   Kitchen Products
*   Fire Protection Equipment
*   Water Meters
*   Carriers and Supports

In addition, this chapter covers labor tables for plumbing specialties such as:

*   Drains: Floor, Roof, Scupper and Gutter types, Floor Cleanouts
*   Back Flow Preventers

Current and local prices for bidding purposes are to be gotten through quotations from suppliers or from a suppliers pricing manual.

Plumbing Fixtures
Labor Only

Lavatories	Labor
Wall Hung	3.26
Comer	4.15
Counter Top	3.05
Sinks	
Counter Top	4.46
Floor	5.95
All Purpose	5.28
Instrument	6.75
Toilets	
One Piece Toilet	3.05
Two Piece Toilet	4.00
Wall Hung	4.76
Bidet	3.05
Urinals	
Wall Hung	4.76
Floor Mounted	5.95
Trough	6.75
Women's w/FV	5.35

Plumbing Fixtures
Labor Only

Drinking Fountains	Labor
Wall Hung	2.96
Wall Recessed	3.27
Freeze Proof	3.78
Free Standing	4.47

Tubs	
Cast Iron Comer	7.48
Cast Iron Recessed	7.48
Cast Iron Square	7.48

Showers	
One Piece Fiberglass Shower and Tub	8.94
32" Square Shower Stall	7.35
36" Square Shower Stall	7.48
36" Comer Shower Stall	7.48

Kitchen Products	
Dishwashing Machine, 45 inch	23.95
Dishwashing Machine, 54 inch	27.90
Dishwashing Machine, 64 inch	32.50
Dishwashing Machine, 80 inch	35.50
Garbage Disposal, 1-1/2 BP	5.95
Garbage Disposal, 5 Bp	9.95
Ice Maker, 225 lbs, 24 hours, Air Cooled	5.95
Ice Maker, 225 lbs, 24 hours, Water Cooled	5.95

Plumbing Fixtures
Labor Only

Fire Protection Equipment	Labor
Fire Hose Cabinet w/Valve & 75' Hose	3.00
Hose Valve with Cap & Chain	2.80

Fire Pump	
500 GPM	50.80
750 GPM	53.35
1000 GPM	62.20
1500 GPM	69.00
2000 GPM	95.00
2500 GPM	104.80
3000 GPM	131.50
3500 GPM	140.00

Fire Extinguishers	
Manual Pull Station	1.25
CO_2 High Pressure System	
75 Lb Cylinder	2.45
100 Lb	2.95
FM200 System	
25 Lb Container	1.85
45 Lb	2.15
65 Lb	2.45
100 Lb	2.95
200 Lb	2.75
Wall Hydrant	4.20

Plumbing Fixtures
Labor Only

Water Meters	Labor
Commercial / Residential—Bronze	
1/2" to 1"	0.55
1-1/4" to 2"	1.08
3"	5.00
4"	9.80
6"	15.05
8"	18.60

Carriers / Supports	
Adjustable Siphon Jet Single Water Closet	
4"-5"	1.80
Adjustable Siphon Jet Double Water Closet	
4"-6"	3.20
Offset Single Blowout Water Closet	
4"	1.80
Offset Double Blowout Water Closet	
4"	3.20
Floor/Wall Mount Urinal	1.80
Lavatory	1.55
Sink	1.55
Drinking Fountain	1.55
Water Coolers	1.55

Plumbing Specialties
Labor Only

Drains	Labor
Floor Drain—Standard	
2" Dia	1.32
3" Dia	1.42
4" Dia	1.58
5" Dia	1.79
6" Dia	2.00
Roof Drain—Standard	
2" Dia	1.58
3" Dia	1.68
4" Dia	1.84
5" Dia	2.10
6" Dia	2.37
Scupper Drain—Standard	
2" - 4" Dia	0.95
5" - 6" Dia	1.10
Gutter Drain—Standard	
2" Dia	1.32
3" Dia	1.42
4" Dia	1.58
5" Dia	1.79
6" Dia	2.00

Plumbing Specialties
Labor Only

Cleanouts	Labor
Floor Cleanout	
2" Dia	0.86
3" Dia	0.92
4" Dia	1.09
5" Dia	1.51
6" Dia	1.70
8" Dia	1.89
Wall Cleanout	
2" Dia	0.50
3" Dia	0.53
4" Dia	0.56
5" Dia	0.59
6" Dia	0.61
8" Dia	0.65

Backflow Preventers	
Threaded Bronze	
3/4"	1.23
1"	1.46
1-1/4"	1.93
1-1/2"	2.16
2"	2.84
Flanged Iron	
2-1/2"	4.00
3"	4.39
4"	5.80
6"	8.52
8"	10.33
10"	12.35

Plumbing Specialties
Labor Only

Backflow Preventers	Labor
Union Bronze	
3/4"	1.23
1"	1.46
Double Check Valve Threaded	
3/4"	0.78
1"	0.95
1-1/2"	1.14
2"	1.39
Double Check Valve Flanged	
2-1/2"	2.41
3"	4.02
4"	5.55
6"	8.07
8"	9.99
1001	12.16

Chapter 8

Air Pollution and
Heat Recovery Equipment

AIR POLLUTION EQUIPMENT

There are four basic types of collectors, each with its own variations. These are: centrifugal cyclones, baghouses, scrubbers, and electrostatic precipitators. Collectors are rated by the size of particles they can remove, by percentage efficiency of removal of different size particles, and by permissible temperature ranges, pressure drops, air volumes, and water usage.

Cyclones are dry centrifugal collectors that remove particles down to approximately 10 microns at efficiencies in the 80 to 90 percent region. Particles are removed by centrifugal force and drop down into a hopper while the cleaned air goes out the top of the unit. There are three variations of cyclones. The first is the large diameter, low resistance type, which is most efficient in the 40 to 60 micron particle size range. The second is the medium efficiency cyclone, which has a smaller diameter housing and is generally used for 20 to 30 micron sized particles. The third cyclone is the high efficiency, small diameter model, which is very efficient in the 10 to 15 micron range.

Baghouses are dry collectors that clean like a vacuum cleaner, incorporating a number of tubular, stocking-like fabric filter bags. They can collect dry particulates such as dust and fumes only and cannot be used for gases or liquids. They are highly efficient, reaching a 99.9 percent level on 1 micron size particles, and operate best in the 0.25 micron size range. Pressure losses are higher, around 3 to 8 inch WG, and are dependent on the air-to-

cloth ratios involved. There are two basic types of fabric collectors, distinguished by the method of bag cleaning and direction or air flow. The mechanical baghouse has motorized shakers to remove dust from the insides of the vertical tubes. The dust laden air flows up and into the tubes, then out through the fabric. The other type of baghouse is pneumatically cleaned-air is used to clean the bags rather then mechanical shakers-and there are three variations: reverse air flow, reverse jet, and pulse jet.

Wet scrubbers can be used for particulates, gases, and liquids. Liquid droplets in a scrubber capture particulates mechanically while gases are removed through absorption. There are a number of different types of scrubbers such as packed towers, venturi, wet centrifugal, wet dynamic, orifice types, and fog towers. Efficiencies range from 90 to 99 percent for particles in the 1 to 5 micron range. Mgh energy scrubbers remove particles in the 0.25 to 1 micron range. Wet scrubbers are often divided into three groups according to pressure losses, theses being 4 to 8, 8 to 25, and 25 to 60 inch WG, respectively.

Electrostatic precipitators are used primarily to collect particulates such as welding fumes and some mists. Efficiencies are in the 98 to 99 percent range for 0.25 to 1 micron particles. A precipitator employs an ionization process whereby incoming particles are made negative and then collected on a positive plate through magnetic attraction. The collector plate is periodically cleaned by rapping, causing the particles to fall away by gravity.

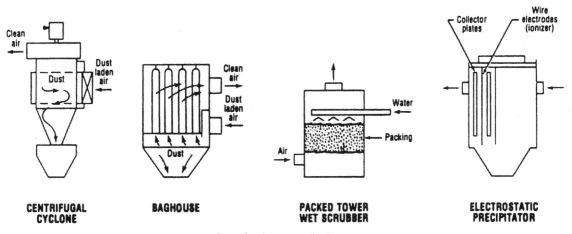

Four basic types of collectors

High Efficiency Centrifugal Cyclone Collectors
Includes Fan

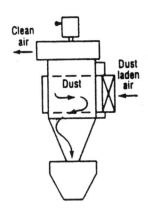

| CFM | DIRECT MATERIAL COST | | LABOR | TOTAL MATERIAL & LABOR | |
	Each	Per CFM	Man Hours	Direct Cost	with 30% O&P
1,000	$2,605	$2.60	8	$2,917	$3,792
3,000	6,216	2.07	10	6,606	8,588
5,000	8,658	1.73	15	9,243	12,016
8,000	11,248	1.41	20	12,028	15,636
10,000	12,876	1.29	25	13,851	18,006
15,000	18,870	1.26	32	20,118	26,153
20,000	24,272	1.21	38	25,754	33,480
25,000	28,860	1.15	44	30,576	39,749
30,000	33,744	1.12	50	35,694	46,402
40,000	43,216	1.08	58	45,478	59,121
50,000	------	------	----	-------	--------
60,000	------	------	----	-------	--------

CORRECTION FACTORS   MATERIAL   LABOR
1.   Medium efficiency cyclone
2.   Low efficiency cyclone
3.   Wet centrifugal cyclone   1.25

Mechanical Bag House Collectors

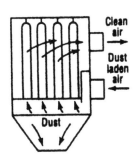

| CFM | DIRECT MATERIAL COST | | LABOR | TOTAL MATERIAL & LABOR | |
	Each	Per CFM	Man Hours	Direct Cost	with 30% O&P
1,000	$5,668	$5.67	28	$6,760	$8,789
3,000	13,986	4.66	34	15,312	19,906
5,000	19,536	3.91	40	21,096	27,425
8,000	24,154	3.02	48	26,026	33,833
10,000	26,788	2.68	56	28,972	37,664
15,000	37,518	2.50	60	39,858	51,815
20,000	45,584	2.28	64	48,080	62,504
25,000	54,390	2.18	68	57,042	74,155
30,000	62,160	2.07	72	64,968	84,458
40,000	82,288	2.06	85	85,603	111,284
50,000	102,860	2.06	96	106,604	138,585
60,000	122,544	2.04	108	126,756	164,783

CORRECTION FACTORS
1.   Pneumatic Bag House 0.90 times value for mechanical type (labor).
2.   Pneumatic Bag House 0.71 material multiplier for pneumatic bag house.

Wet Scrubber Collectors

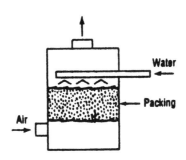

CFM	DIRECT MATERIAL COST		LABOR	TOTAL MATERIAL & LABOR	
	Each	Per CFM	Man Hours	Direct Cost	with 30% O&P
1,000			22	$1,270	$1,651
3,000	12,965	4.32	28	14,057	18,274
5,000	16,798	3.36	32	18,046	23,460
8,000	20,838	2.60	38	22,320	29,017
10,000	23,532	2.35	45	25,287	32,873
15,000	32,190	2.15	48	34,062	44,281
20,000	39,072	1.95	54	41,178	53,531
25,000	45,880	1.84	58	48,142	62,585
30,000	53,724	1.79	62	56,142	72,985
40,000	69,856	1.75	68	72,508	94,260
50,000	85,100	1.70	76	88,064	114,483
60,000	99,456	1.66	88	102,888	133,754

Electrostatic Precipitator Collectors

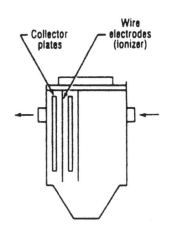

CFM	DIRECT MATERIAL COST		LABOR	TOTAL MATERIAL & LABOR	
	Each	Per CFM	Man Hours	Direct Cost	with 30% O&P
1,000	$3,078	$3.08	8	$3,390	$4,408
3,000	8,036	2.68	10	8,426	10,954
5,000	12,062	2.41	15	12,647	16,441
8,000	17,168	2.15	20	17,948	23,332
10,000	19,980	2.00	25	20,955	27,242
15,000	28,416	1.89	32	29,664	38,563
20,000	35,816	1.79	38	37,298	48,487
25,000	43,660	1.75	44	45,376	58,989
30,000	51,060	1.70	50	53,010	68,913
40,000	65,712	1.64	58	67,974	88,366
50,000	80,660	1.61	66	83,234	108,204
60,000	96,792	1.61	74	99,678	129,581

Industrial Exhaust Fans

CFM	DIRECT MATERIAL COST		LABOR	TOTAL MATERIAL & LABOR	
	Each	Per CFM	Man Hours	Direct Cost	with 30% O&P
1,000	$1,672	$1.67	4.00	$1,828	$2,377
2,000	3,078	1.54	9.00	3,429	4,458
4,000	3,907	0.98	13.00	4,414	5,738
6,000	5,062	0.84	16.00	5,686	7,391
8,000	6,512	0.81	18.00	7,214	9,378
10,000	7,844	0.78	21.00	8,663	11,262
12,000	9,413	0.78	23.00	10,310	13,403
14,000	10,982	0.78	26.00	11,996	15,594
16,000	12,077	0.75	29.00	13,208	17,170
18,000	13,586	0.75	31.00	14,795	19,234
20,000	13,616	0.68	34.00	14,942	19,425
24,000	16,339	0.68	39.00	17,860	23,218
28,000	19,062	0.68	44.00	20,778	27,012
32,000	21,786	0.68	49.00	23,697	30,806
36,000	23,443	0.65	55.00	25,588	33,265
40,000	26,048	0.65	60.00	28,388	36,904

CORRECTION FACTORS
1. For suspended fan, +20%.
2. For DWDI fan, +10%
3. For split centrifugal housing, knocked down, +50%.
4. For inlet vane control, +10%.
5. For crane hoist, with fan set directly on pad, -20%.

Gas Fired Makeup Air Units

CFM	DIRECT MATERIAL COST		LABOR	TOTAL MATERIAL & LABOR	
	Each	Per CFM	Man Hours	Direct Cost	with 30% O&P
1,000	$1,051	$1.05	6	$1,285	$1,670
2,000	2,072	1.04	12	2,540	3,302
4,000	4,144	1.04	15	4,729	6,148
6,000	6,038	1.01	20	6,818	8,864
8,000	7,933	0.99	22	8,791	11,428
10,000	9,768	0.98	24	10,704	13,915
12,000	11,544	0.96	26	12,558	16,325
14,000	13,261	0.95	29	14,392	18,709
16,000	15,155	0.95	31	16,364	21,273
18,000	16,783	0.93	34	18,109	23,542
20,000	18,352	0.92	36	19,756	25,683
24,000	21,667	0.90	37	23,110	30,043
28,000	25,278	0.90	38	26,760	34,789
32,000	28,416	0.89	42	30,054	39,070
36,000	31,968	0.89	46	33,762	43,891
40,000	34,928	0.87	50	36,878	47,941

Air to Air Plate Heat Exchangers

CFM	DIRECT MATERIAL COST		LABOR	TOTAL MATERIAL & LABOR	
	Each	Per CFM	Man Hours	Direct Costs	with 30% O&P
5,000	$8,140	$1.63	12	$8,608	$11,190
10,000	14,252	1.43	15	14,837	19,289
15,000	18,204	1.21	20	18,984	24,679
20,000	19,536	0.98	24	20,472	26,614
25,000	22,200	0.89	25	23,175	30,128

ACCESSORIES
1. Face and by-pass section with controls
    5,000 CFM $2,516
    10,000 CFM $2,960

Direct labor costs are $39.00 per hour.

Heat Exchanger Wheels

CFM	DIRECT MATERIAL COST		LABOR	TOTAL MATERIAL & LABOR	
	Each	Per CFM	Man Hours	Direct Costs	with 30% O&P
PACKAGE UNIT					
7,000	$51,800	$7.40	15	$52,502	$68,253
10,000	64,380	6.44	26	65,394	85,012
15,000	75,480	5.03	32	76,728	99,746
20,000	81,400	4.07	40	82,960	107,848
25,000	83,250	3.33	41	84,849	110,304
30,000	93,240	3.11	42	94,878	123,341
35,000	106,190	3.03	51	108,179	140,633
40,000	118,400	2.96	55	120,545	156,709
50,000	148,000	2.96	60	150,340	195,442
WHEEL ONLY					
7,000	$12,284	$1.75	12	$12,752	$16,578
10,000	16,280	1.63	15	16,865	21,925
15,000	22,200	1.48	20	22,980	29,874
20,000	23,976	1.20	24	24,912	32,386
25,000	24,420	0.98	25	25,395	33,014
30,000	25,752	0.86	26	26,766	34,796
35,000	26,936	0.77	31	28,145	36,589
40,000	29,008	0.73	33	30,295	39,384
50,000	35,520	0.71	36	36,924	48,001

Package unit includes 2 fans, AC inverter on motor for wheel, all controls.

Direct labor costs are $39.00 per hour.

Exhaust Gas Heat Exchanger Shells
Carbon Steel

DIA.	LENGTH	DIRECT COST	LABOR	TOTAL MATERIAL & LABOR	
Inches	Inches	Each	Man Hours	Direct Costs	with 30% O&P
4	26	2,950	3.6	$3,100	$4,031
6	34	4,440	7.5	4,733	6,152
8	42	5,920	9.0	6,271	8,152
12	58	10,360	11.4	10,805	14,046
16	74	16,280	17.5	16,963	22,051
20	90	25,160	21.0	25,979	33,773
24	106	37,000	25.5	37,995	49,393
28	127	47,360	32.0	48,608	63,190

CORRECTION FACTORS
1. Stainless steel, steel material, plus 50%

ACCESSORIES
1. Manual Diverter $1,500 to $3,500
2. Automatic Diverter $1,865 to $4,000
3. Removable Manifold $400 to $1,000

Paybacks, 1 to 2 years
Heat taken from engine exhaust, etc.

USES OF RECOVERED HEAT
1. Heating air for HVAC systems.
2. Use of evaporation cycle of AC system.
3. Preheating feedwater to boilers.

Direct labor costs are $39.00 per hour.

Section IV

Sheet Metal Estimating

This page intentionally left blank

Chapter 9

Sheet Metal Estimating Basics

REQUIREMENTS OF A
PROFICIENT SHEET METAL ESTIMATOR

The crux of successful contracting is built on a foundation of complete and accurate estimates with proper markups.

Solid estimates are produced by competent and reliable estimators. Good sheet metal estimators are developed through the following background of knowledge, procedures, skills and abilities:

ESTIMATING PRINCIPLES AND PROCEDURES

1. They must follow sound efficient procedures for preparing estimates, such as:

- Become thoroughly familiar with the project, the types of systems and ductwork involved, in the scope of work, etc. before starting detailed takeoffs.

- Be Familiar with budget estimating; HVAC costs for different buildings based on cost per square foot of building or cost per ton of air conditioning, amount of ductwork per square foot of building or by the average size, cost of ductwork per linear foot, per pound or per square foot.

- Know the major categories of an estimate:
  Equipment
  Ductwork
  Piping
  Duct Accessories and Sheet Metal Specialties
  Special Labor
  Sub-Contractors
  End of Bid Factors (such as sales tax)
  Markups for Overhead and Profit

- Must be familiar with detailed scope of what is required in a sheet metal estimate.

- Highlight drawings before doing the takeoff.

- Follow systematic overall procedure.
  Study the plans and specs
  Send out quotation requests
  Highlight Drawings
  Make Takeoff; and Extensions

Summarize
Recap and markups

- Do constant systematic checking on each part as you go along and overall at the end. Double check everything.

2. They must have the ability to read blue prints, recognize symbols, types of ductwork, equipment and systems, etc.

Air Distribution Systems

3. They must be familiar with the different types of HVAC systems such as:
  Low pressure constant volume systems
      Single zone, reheat coils, multi-zone
  High pressure constant volume systems
      Dual duct, induction, reheat terminals
  Variable air volume
      Cooling only, cooling/reheat terminals
      Fan powered, dual duct
      Induction, multi-zone
      System powered, riding fan curve
      Damper terminal by-pass
  Exhaust systems
      Return air, toilet exhaust
      Kitchen, lab, industrial

They must not only recognize the various types of systems on plans, but they must know all of the components required in them, whether shown on plans or not.

4. They must know duct pressure and velocity ranges:
  Duct Pressure Ranges:
      1/2 inch, 1 inch, 2 inch static pressure
      3, 4, 5 inch static pressure
      6 inch static pressure and up
  Velocities: 0 to 2,000 fpm
              2,000 fpm and up, etc.

5. They must know about different ductwork system configurations such as:
      Single Duct

Dual Duct
Multi-Zone
Loops
Plenum Ceilings

6. They must have some familiarity with air distribution system design, know the recommended air speeds, pressure drops and duct sizing and selection of equipment.

Types of Ductwork

7. A sheet metal estimator must be familiar of the different types of ductwork and their correct construction.

    Rectangular galvanized: Low, medium and high pressure

    Low pressure round ductwork; flues, flexible tubing

    Spiral pipe and fittings

    Light gauge aluminum, stainless, PVC with cleats, pittsburghs

    Heavy gauge metals; black iron, stainless, aluminum, galvanized, corton, etc.

    PVC, FRP, Sundstrand

8. They must know the correct applications of different types of ductwork materials to various systems:
    Low, medium, high pressure HVAC systems
    General exhausts
    Fume exhausts
    Heat exhaust systems
    Chemical exhaust systems
    Abrasive material systems

Ductwork Construction

9. They must be familiar with the different type of connections for each type of ductwork and their correct application to different types of systems.
    Cleats; drive, flat S, standing S, bar, reinforced bar
    Transverse; TDC, TDF
    4 bolt connections
    Angle flange, vane stone
    Bent angle flange
    Butt welded
    Slip, couplings

10. They must be familiar with different types of seams used for constructing ductwork.
    Pittsburgh
    Snaplock, lockseam
    Welded

11. They must be familiar with the different gauges used for ductwork and specialties.
    Commercial galvanized 26 through 16 gauge
    Residential galvanized 3 0 through 18 gauge
    Heavy gauge industrial 18 gauge through 1/2 inch thick plates
    Fiberglass ductboard, 1 inch thick
    PVC, 1/4, 3/16 inch thick

12. They must be familiar with the different types of reinforcing used on ductwork.
    Angles
    Channels
    Cross breaking
    Tie rods

13. They must be familiar with all the different types of fittings used in air distribution systems and of their correct application.
    Elbows; 90*, 45*, 22-1/2*, etc.
        Radius throat, square throat
    Transitions; equal taper, FOT, FOB, square side, etc.
    Offsets; Ogee, square
    Wye fittings
    Tap in tees

Estimating Materials and Labor

14. They must know the various methods of estimating ductwork.

15. A good sheet metal estimator must know how to estimate ductwork materials.
    Takeoff and calculate surface square footage of material based on size, length, etc.
    Add waste and seam factors
    Multiply by weight per square foot

16. They must be familiar with different waste and allowance factors for seams, cleats, hangers, hardware, etc.

17. They must know the methods of estimating ductwork labor such as:

Per Piece	Per Breakdown of
Per Pound	component parts
Per Square Foot	Per Linear Foot
Per Batch	

They must know sources of labor such as the Wendes Mechanical Estimating Manual, cost records, etc.

Correction Factors

18. They must apply labor multipliers with reasonable accuracy whenever needed to adjust for conditions, such as:

    5th floor takes 10 percent longer

    30 foot high ductwork takes 20 percent longer

    Duplicate fittings go 33 percent faster

Accessories

19. They must be familiar with the various duct accessories and sheet metal specialties.

    Turning vane's air foil, single skin

    Splitter dampers

    Canvas or flex connections

    Single and multiblade dampers

    Access doors

Fabrication and Installation Procedures

20. They must be familiar with fabrication procedures and machinery and how they affect labor and overhead margins. They must be familiar with plasma cutters, coil lines, seam machines, press breaks, rollers, etc.

    Plasma cutters cut overall fitting labor in half

    Duct coil lines reduce straight duct labor by about 70 percent

21. A Sheet metal estimator must be well versed in ductwork installation procedures, in the operations involved in installations, with hand tools, scaffolding, vermets, scissor hoists, etc.

Pricing Equipment

22. They must know sources of pricing on accessories and equipment, suppliers, price catalogues, suppliers for quotations, etc.

23. They have to know about small ventilation equipment.

    Grilles, registers, diffusers

    Multiblade dampers, back draft dampers

    Fire dampers, access doors

24. They must know about sheet metal specialties such as:

    Sheet metal housings, walk through doors

    Belt guards, drain pans, coil stands

    Coil blank offs

25. They must know about major HVAC equipment.

    Roof top units, air handling units

    Fans, filters, louvers

Wage Rates, Unions, Jurisdictions

26. They must know about wage rates, union fringe benefits, federal, state and local taxes, insurance's, etc.

27. They must be knowledgeable about union, trade and local labor jurisdictions.

28. They must be familiar with building codes.

Other Trades, Types of Buildings

29. They have to be familiar with other trades such as piping, insulation, temperature control, electrical and excavation.

30. They must be familiar with all types of buildings, commercial, institutional, their general sizes, layout, etc. and with the sequence of general construction work.

Markups

31. A good sheet metal estimator must be generally familiar with financial statements such as profit loss and balance sheets. They must be able to determine the correct markup for overhead and profit for their company and for the particular job they are bidding.

    They should understand how overhead costs are pro-rated onto direct material and labor costs for different projects, for different levels of sales and overhead costs, for different ratios of material to labor, etc.

Skills, Traits Required

32. Estimating requires a host of skills, mathematical, mechanical, reading, writing, visualizing and drawing. It requires being methodical, analytical, strategically and realistic.

33. It absolutely demands that the estimator be reliable, that they be thorough in their understanding of the project, of it's scope, in takeoffs, interpretations, extensions, summaries and recaps.

    Thus, the knowledgeable, proficient and reliable estimator as described above will be able to produce complete and accurate estimates, which in turn become the required foundation blocks of successful contracting.

TYPES OF DUCTWORK

HVAC Rectangular

1. **Low pressure galvanized** ductwork comprises the

bulk of HVAC ductwork used in buildings. It's used for system pressures between 0-2" S.P. and air velocities between 0-2500 FPM. Generally connections are with cleats and the seams are snaplock or pittsburgh. Reinforcing is either crossbreaking, beading, reinforced cleats or structural angles.

2. **Medium pressure galvanized** ductwork is used for pressures from 2-6" S.P. and velocities from 2000 to 4000 within the systems where the S.P. is over 6" and the velocities are over 2000 FPM.

   Both medium and high pressure ductwork must be sealed to maintain pressures within 1 or 1/2% of design CFM. Both are constructed with pittsburgh seams and the connections are with cleats which you can seal, or are gasketed companion angles. Reinforcing is with angles, either backup near the connection and/or at prescribed intervals.

3. **Fiberglass ductboard** is used for about 15% of all the HVAC ductwork in the U.S. It's primarily used for low pressure systems in unconditioned spaces where insulation is needed or for ductwork that requires acoustic insulation. It's easier to fabricate and install than the galvanized. Boards are one inch thick, seams and connections shiplap grooved, and stapled and taped.

4. **Aluminum** is used for supply ductwork in HVAC systems if exposed to moisture as in a pool area, shower room etc. It is fabricated the same way as low pressure galvanized with pittsburghs, cleats, angles, etc.

HVAC Round

1. **Round galvanized spiral pipe** and fittings is used primarily for high pressure systems with cemented and taped connections but it can also be used for low velocity situations as well as for various types of exhaust systems. The installed cost is slightly less than rectangular galvanized.

2. **Round residential lock seam** ductwork, also known furnace pipe or just galvanized pipe, is most commonly used in residences and apartment buildings as well as for flues. Connections are crimped on one side, slipped together and screwed. Elbows are adjustable.

3. **HVAC flexible tubing**, single skin or factory insulated is used for residential work and for commer-

cial low pressure and high pressure systems.

4. Round **flues**, either single or double skin used for furnaces, unit heaters etc. is the last of the HVAC round ductwork.

Industrial Exhaust Ductwork

There are four basic categories of materials used for ductwork in air pollution control and industrial and commercial exhaust systems.

1. **Galvanized** ductwork, either round or rectangular, is used in many applications for heat, moisture, and dust removal when corrosions and abrasion do not present problems.

2. **Black iron** ductwork, hot or cold rolled, round or rectangular is used in many applications for heat, moisture, and dust removal when corrosion and abrasion do not present problems.

3. **Corrosion and moisture resistant** ductwork, round or rectangular, is fabricated from the following materials: PVC, FRP stainless steel, asbestos-cement, PVC coated galvanized, and aluminum.

4. High temperature ductwork is generally **stainless steel**.

   Round ductwork is used more commonly than rectangular in air pollution control and material conveying work. Connections may be cleated, slip, sheet metal flanged or companion angle flanged and bolted, welded, soldered, cemented, gasket, sleeved, coupled, or riveted.

PROCEDURE FOR TAKING OFF DUCTWORK

1. Study specs, drawings, duct routing, fittings, risers, offsets, materials, connections and hanging requirements first.

2. Mark the different requirements on the drawings. Color runs as required.

3. There are two general procedures for taking off ductwork: one is to takeoff the galvanized ductwork drawing by drawing, and the special rectangular and round ducts, system by system. This generally works well for medium or large size jobs with more than two or three ductwork drawings. Some estimators

Types of Ductwork Connections

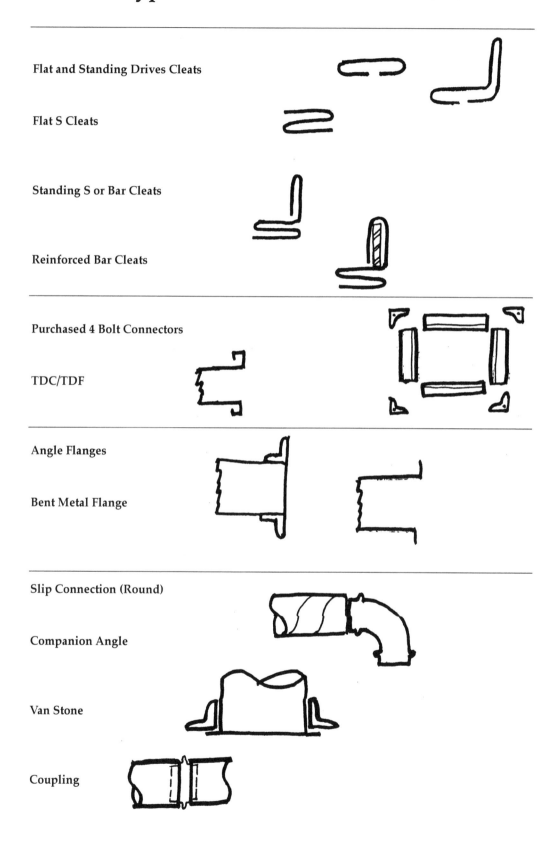

Flat and Standing Drives Cleats

Flat S Cleats

Standing S or Bar Cleats

Reinforced Bar Cleats

Purchased 4 Bolt Connectors

TDC/TDF

Angle Flanges

Bent Metal Flange

Slip Connection (Round)

Companion Angle

Van Stone

Coupling

prefer this approach for all jobs whether small with one drawing, or very large with many drawings.

The other general procedure is to takeoff everything "system by system," both the galvanized and the special round ducts. This is helpful when there is a great deal of congestion on the plans which can cause errors. Each system is taken off completely from beginning to end regardless of how many drawings it spans.

4. The general sequence for taking off the different types of ductwork, whether you do so drawing by drawing, or system by system, is as follows:

- Make sure you have all the lined ducts, alternate areas and correction factor areas taken off first or clearly marked.

- Follow up with the bare galvanized.

- Do the low pressure runs first.

- Then the high pressure ductwork.

- Follow up with heavy gauge industrial ductwork

- Lastly take off the round ductwork.

5. Takeoff ductwork segment by duct segment picking up all connecting branches as you go along, and complete each segment before going to the next.

6. Another approach on complicated duct runs is to take off fittings first to get familiar with the duct runs and to identify accessories etc. within the duct runs so as they are not missed. Then takeoff straight pipe.

7. Identify takeoff sheets with drawing numbers, systems, floors, type of ductwork, connection type, etc. such as "M1, S2, LP Galvanized, Lined" etc. and check off or draw a line through ducts on the drawings as they are taken off.

METHODS OF FIGURING DUCTWORK WEIGHT

1. Weight Per Running Foot
This is the traditional way of arriving at material weight for HVAC galvanized and other types of metal ductwork.

Long Hand Method
The sq ft/ft of ductwork if multiplied times the linear feet involved, then multiplied by the weight per sq ft of metal for that gauge, and finally a 20% allowance is added for waste, hangers, cleats, seams, etc.

Combining Pound Per Sq Ft and Allowance into One Factor
The factors 1. 156 lbs/SF and 1.2 allowance can be combined into a single multiplying factor of 1. 156 x 1.2 = 1.4. Hence 60 SF x 1.4 = 84 lbs.

Duct Size	Sq Ft/Ft	Length	Square Feet
24 x 12	6	10 Ft	60 SF Total
			x <u>1.156</u> Lbs/Sq Ft
			69.4 Lbs Raw
			x <u>1.2</u> + 20% allow
			84 Lbs total

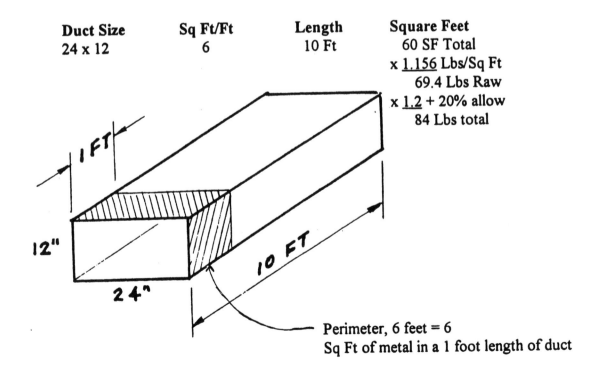

Perimeter, 6 feet = 6
Sq Ft of metal in a 1 foot length of duct

Using Chart to Read Pounds Per Foot Directly

Using precalculated Lbs/Ft from chart eliminates a great deal of wasted repetitive math, calculations and writing.

When using the chart you simply determine and locate the semi-perimeter of the duct, read the weight per foot on the chart according to the gauge (which already has the 20% allowance built into it) and multiply it times the length of duct.

		From Chart Lbs/Ft w/20%	
Duct Size	Length	Waste Built In	Total Weight
24 x 12	10 Ft	8.4	84 lbs

This is a much faster and simpler method for calculating duct weight per foot than the long hand method, converting to sq ft for lined and insulated ductwork. If the duct run happens to be lined or insulated you simply divide the total weight by whatever lbs/ft of metal you used to start with, and you will be converted to square feet.

Example: 84 lbs of 24 gauge—1.4 lbs/sq ft = 60 sq ft

2. Square Feet Per Foot

For non-metallic ducts such as fiberglass ductboard, plastic PVC, fiberglass reinforced plastic FRP, lining and insulation, the traditional method for determining the material required for ductwork is similar to the pounds per foot for metal ducts, except that one factor is left out, the weight per sq ft.

Example:
Using the same duct as in example from above:

Duct Size	Sq Ft/Ft	Length	Square Feet
24 x 12	6	10 Ft	6.0
			x 1.2 Waste Factor
			7.2 Sq Ft Gross

3. Weight Per Piece

Actual weight per piece is used based on prior weighing, or from the actual recorded amount of material used for the particular item.
* 42 x 18 transition, 3 foot long weighs 51 lbs
* 18 x 9 joint of pie, 5 foot long weighs 32 lbs

4. Actual Sheets Needed

Material is determined by actual sheets of metal needed to fabricate the ductwork. This is normally done for special, more expensive materials, for odd configurations, or for smaller projects.

METHODS OF CALCULATING DUCTWORK LABOR

1. Hours Per Piece

The actual amount of labor to fabricate or install each specific piece is predetermined and applied to the various sizes and types of ductwork to arrive at the number of hours needed. (See figure on following page.)

2. Pounds Per Hour (or hours per pound)

A pounds per hour productivity measurement is used to determine labor.

Shop rate for typical LP galvanized is about 44 lbs/hr

Installation rate about 25 lbs/hr

Or lbs/hr can be converted to "hours per pound."

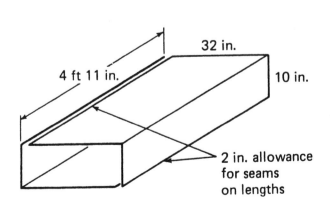

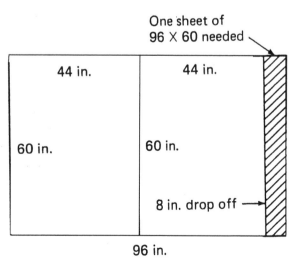

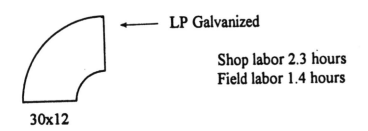

LP Galvanized

Shop labor 2.3 hours
Field labor 1.4 hours

30x12

44 lbs/hr divided into 1 = .023 hrs/lb
25 lbs/hr is the same as .04 hrs/lb

3.   **Square Foot Per Hour**
Many special material ducts are calculated on a. "SF/W' basis for labor, fiberglass, PVC, and FRP being some.

Square feet per hour is an excellent way to compare costs of different gauges of ductwork and of different types of ductwork. It does away with many of the confusing variables involved with weight and gives you a more direct comparison.

Fiberglass ducts fabricate at a rate of 55 SF/hr and install at 30 SF/hr

If you converted typical 24 gauge LP galvanized ductwork productivity rates into square feet you would get:

$$\frac{44\,lbs/hr}{1.156} = 38\,SF/hr \qquad \frac{25\,lbs/hr}{1.156} = 22\,SR/hr$$

4.   **Man Days**
The time needed to fabricate or install ductwork or equipment is calculated in terms of days it would take one man or a crew to perform the work.

5.   **Per Operation**
The time needed to perform each operation, such as layout, cutting, bending, assembly, etc., is calculated separately and then added together.

20% Allowance Factor for Galvanized Ductwork

A 20% allowance must be added to the surface area of ductwork to cover hangers, cleats, hardware, waste and seams.

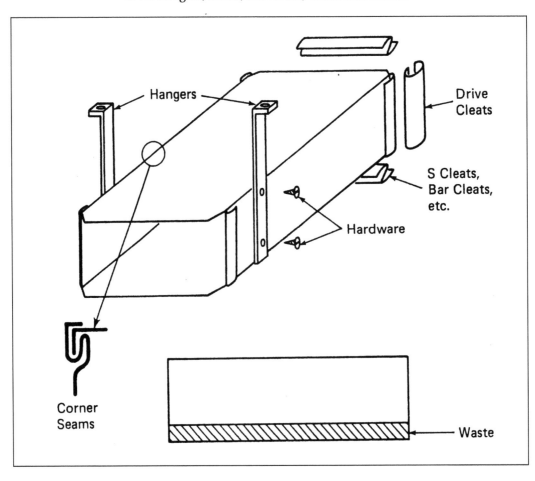

Weight of Galvanized Ductwork Per Linear Foot
With 20% Allowance

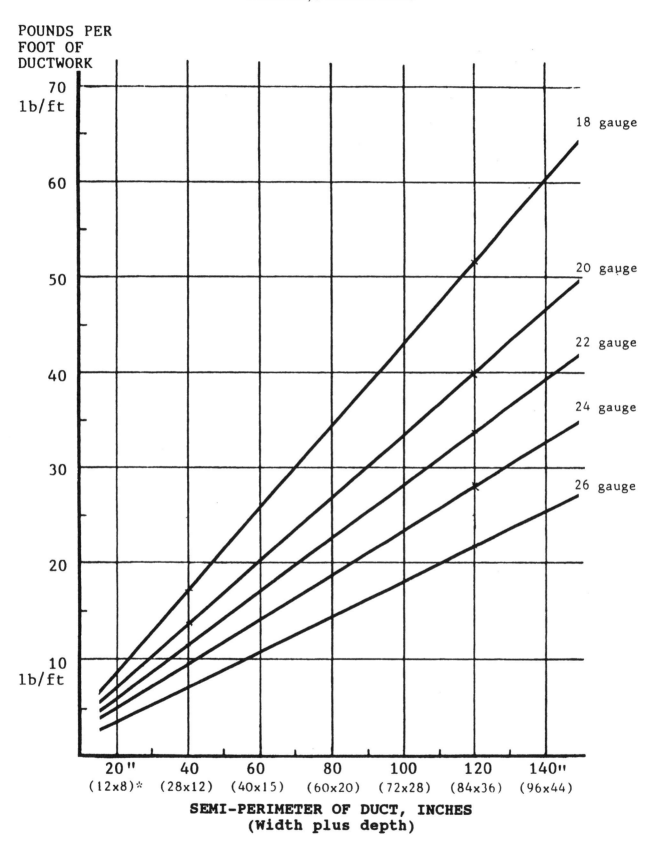

POUNDS PER
FOOT OF
DUCTWORK

SEMI-PERIMETER OF DUCT, INCHES
(Width plus depth)

* Sample size

Weight of Galvanized Ductwork Per Linear Foot
With 20% Allowance

SEMI-PERIM	0-12	13-30	31-54	55-84	85 up	Square Feet per LF (no allow.)
	26 ga	24 ga	22 ga	20 ga	18 ga	
Width + Depth	.91 lbs/SF 1.10 W/20%	1.16 1.40	1.41 1.70	1.66 2.00	2.16 2.60	
12	2.20	2.80	3.40	4.00	5.2	2.00
14	2.58	3.28	3.98	4.68	6.09	2.34
16	2.94	3.74	4.54	5.34	6.95	2.67
18	3.30	4.30	4.2	5.10	6.00	3.00
20	3.68	4.68	5.68	6.68	8.69	3.34
22	4.04	5.14	6.24	7.34	9.55	3.67
24	4.40	5.60	6.80	8.00	10.4	4.00
26		6.08	7.38	8.68	11.29	4.34
28		6.54	7.94	9.34	12.15	4.67
30		7.02	8.50	10.00	13.00	5.00
32		7.48	9.08	10.68	13.85	5.34
34		7.94	9.68	11.34	14.75	5.67
36		8.40	10.10	12.00	15.6	6.00
38		8.88	10.88	12.68	16.45	6.34
40		9.34	11.32	13.34	17.34	6.67
42		9.8	11.90	14.00	18.20	7.00
44		10.28	12.48	14.68	19.24	7.34
46		10.74	13.00	15.34	20.00	7.67
48		11.20	13.60	16.00	20.80	8.00
50		11.68	14.27	16.68	21.69	8.34
52		12.14	14.93	17.34	22.55	8.67
54		12.6	15.30	18.00	23.40	9.00
56		13.00	15.87	18.68	24.29	9.34
58		13.54	16.43	19.34	25.15	9.67
60		14.00	17.00	20.00	26.00	10.00
62			17.57	20.68	26.87	10.34
64			18.13	21.34	27.73	10.67
66			18.70	22.00	28.6	11.00
68			19.27	22.68	29.47	11.34
70			19.83	23.34	30.33	11.67
72			20.40	24.00	31.20	12.00
74			20.98	24.68	32.07	12.34
76			21.55	25.34	32.93	12.67
78			22.10	26.00	33.80	13.00
80			22.67	26.68	34.67	13.34
82			23.23	27.34	35.53	13.67
84			23.80	28.00	36.40	14.00
86			24.37	28.68	37.27	14.34
88			24.93	29.34	38.13	14.67
90			25.50	30.00	32.00	15.00
92			26.07	30.68	40.00	15.34
94			26.63	31.34	40.87	14.67
96			27.20	32.00	41.60	16.00
98			27.77	32.68	42.47	16.34
100			28.33	33.34	43.31	16 67
102			28.90	34.00	44.20	17.00
104			29.47	34.68	45.09	17.34
106			30.03	35.34	46.96	17.67
108			30.60	36.00	46.80	18.00
110				36.68	47.67	18.34
112				37.34	48.53	18.67
114				38.00	49.40	19.00
116				38.68	50.27	19.34
118				39.34	51.13	19.67
120				40.00	52.00	26.00

Weight of Hot Rolled Steel Ductwork Per Linear Foot
No Allowance Included

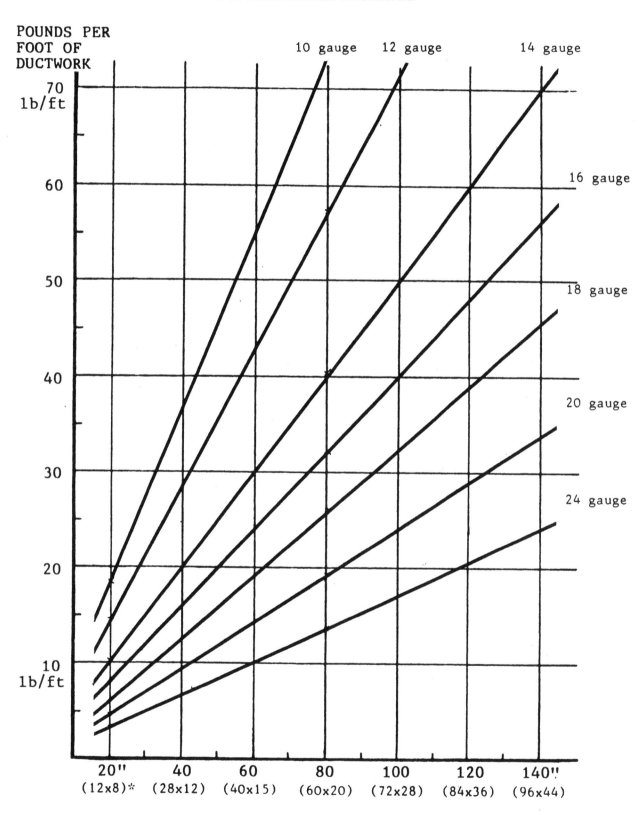

SEMI-PERIMETER OF DUCT, INCHES
(Width plus depth)

* Sample size

Weight of Rectangular Black Iron Ductwork
Per Linear Foot
No Allowance Included, 26 Through 12 Gauge

Semi -Perim				Gauges and Thicknesses					
In Inches W+D	26 0.0179	24 0.0239	22 0.0299	20 0.0359	18 0.0478	16 0.0598	14 0.0747	12 0.1046	SQ FT LF
12	1.46	1.95	2.44	2.92	3.90	4.88	6.10	8.54	2.00
14	1.71	2.28	2.85	3.42	4.57	5.71	7.14	10.00	2.33
16	1.95	2.60	3.25	3.90	5.21	6.51	8.14	11.40	2.67
18	2.19	2.93	3.66	4.38	5.85	7.32	9.14	12.80	3.00
20	2.44	3.25	4.06	4.88	6.52	8.15	10.19	14.26	3.33
22	2.68	3.58	4.48	5.36	7.16	8.95	11.19	15.67	3.67
24	2.92	3.90	4.88	5.84	7.80	9.76	12.19	17.07	4.00
26	3.17	4.23	5.29	6.34	8.47	10.59	13.23	18.53	4.33
28	3.41	4.55	5.69	6.82	9.11	11.39	14.24	19.93	4.67
30	3.65	4.88	6.10	7.30	9.75	12.19	15.24	21.34	5.00
32	3.90	5.20	6.50	7.80	10.42	13.03	16.28	22.80	5.33
34	4.14	5.53	6.91	8.28	11.06	13.83	17.28	24.20	5.67
36	4.38	5.85	7.31	8.76	11.70	14.63	18.29	25.61	6.00
38	4.63	6.18	7.73	9.26	12.37	15.47	19.33	27.07	6.33
40	4.87	6.50	8.13	9.74	13.01	16.27	20.33	28.47	6.67
42	5.12	6.83	8.54	10.24	13.68	17.11	21.38	29.0	7,00
44	5.36	7.16	8.95	10.72	14.32	17.91	22.38	31.33	7.33
46	5.60	7.48	9.35	11.20	14.96	18.71	23.38	32.74	7.67
48	5.85	7.81	9.76	11.70	15.63	19,54	24.42	34.20	8.00
50	6.09	8.13	10.16	12.18	16.27	20.35	25.43	35.60	8.33
52	6.33	8.46	10.58	12.66	16.91	21.15	26.43	37.01	8.67
54	6.58	8.78	10.98	13.16	17.58	21.98	27.47	38.47	9.00
56	6.82	9.11	11.39	13.64	18.22	22.79	28.47	39.87	9.33
58	7.06	9.43	11.79	14.12	18.86	23.59	29.48	41.27	9.67
60	7.31	9.76	12.20	14.62	19.53	24.42	30.52	42.73	10.00
62	7.55	10.08	12.60	15.10	20.17	25.22	31.52	44.14	10.33
64	7.79	10.41	13.01	15.58	20.81	26.03	32.52	45.54	10.67
66	8.04	10.73	13.41	16.08	.21.47	26.86	33.57	47.00	11.00
68	8.28	11.06	13.83	16.56	22.12	27.66	34.57	48.40	11.33
70	8.53	11.38	14.23	17.06	22.78	28.50	35.61	49.87	11.67
72	8.77	11.71	14.64	17.54	23.42	29.30	36.61	51.27	12.00
74	9.01	12.03	15.04	18.02	24.07	30.10	37.62	52.67	12.33
76	9.26	12.36	15.45	18.52	24.73	30.94	38.66	54.13	12.67
78	9.50	12.68	15.85	19.00	25.37	31.74	39.66	55.54	13.00
80	9.74	13.01	16.26	19.48	26.02	32.54	40.66	56.94	13.33
82	9.99	13.33	16,66	19.98	26.68	33.38	41.71	58.40	13.67
84	10.23	13.66	17.08	20.46	27.32	34.18	42.71	59.80	14.00
86	10.47	13.98	17.48	20.94	27.97	34.98	43.71	61.21	14.33
88	10.72	14.31	17.89	21.44	28.63	35.82	44.76	62.67	14.67
90	10.96	14.64	18.30	21.92	29.27	36.62	45.76	64.07	15.00

Weight of Rectangular Hot Rolled Ductwork Per Linear Foot
No Allowance Included, 10 Gauge Through 1/2″ Plate

Semi -Perim In Inches W+D	10 0.135	8 0.164	7 0.179	3/16 0.1876	1/4 0.25	5/16 0.3125	3/8 0.375	1/2 0.5	SQFT LF
12	11.02	13.39	14.61	15.31	20.41	25.51	10.61	40.82	2.00
14	12.86	15.62	17.05	17.86	23.82	29.77	35.73	47.63	2.33
16	14.70	17.86	19.49	20.42	27.22	34.03	40.84	54.45	2.67
18	16.53	20.08	21.91	22.96	30.61	38.26	45.92	61.23	3.00
20	18.37	22.32	24.35	25.52	34.02	42.52	51.03	68.04	3.33
22	20.21	24.56	26.79	28.07	37.43	46.78	56.14	74.86	3.67
24	22.04	26.78	29.22	30.61	40.82	51.02	61.23	81.64	4.00
26	23.88	29.01	31.66	33.17	44.23	55.28	66.34	88.45	4.33
28	25.72	31.25	34.10	35.73	47.63	59.54	71.45	95.27	4.67
30	27.56	33.49	36.54	38.28	51.04	63.80	76.56	102.08	5.00
32	29.39	35.71	38.96	40.82	54.43	68.03	81.65	108.86	5.33
34	31.23	37.94	41.40	43.38	57.84	72.29	86.76	115.68	5.67
36	33.07	40.18	43.84	45.93	61.25	76.55	91.87	122.49	6.00
38	34.90	42.40	46.27	48.48	64.63	80.79	96.95	129.27	6.33
40	36.74	44.64	48.71	51.03	68.04	85.05	102.06	136.08	6.67
42	38.58	46.87	51.15	53.59	71.45	89.30	107.18	142.90	7.00
44	40.42	49.11	53.58	56.14	74.86	93.56	112.29	149.72	7.33
46	42.25	51.33	56.01	58.69	78.25	97.80	117.37	156.49	7.67
48	44.09	53.57	58.45	61.24	81.65	102.06	122.48	163.31	8.00
50	45.93	55.80	60.89	63.80	85.06	106.32	127.59	170.12	8.33
52	47.76	58.03	63.32	66.34	88.45	110.55	132.68	176.90	8.67
54	49.60	60.26	65.75	68.89	91.86	114.81	137.79	183.72	9.00
56	51.44	62.50	68.19	71.45	95.27	119.07	142.90	190.53	9.33
58	53.28	64.74	70.63	74.01	98.67	123.33	148.01	197.35	9.67
60	55.11	66.96	73.06	76.55	102.06	127.57	153.10	204.13	10.00
62	56.95	69.19	75.50	79.10	105.47	131.83	158.21	210.94	10.33
64	58.79	71.43	77.94	81.66	108.88	136.09	163.32	217.76	10.67
66	60.62	73.65	80.36	84.20	112.27	140.32	168.40	224.54	11.00
68	62.46	75.89	82.80	86.76	115.68	144.58	173.51	231.35	11.33
70	64.30	78.12	85.24	89.31	119.08	148.84	178.63	238.17	11.67
72	66.13	80.35	87.67	91.85	122.47	153.08	183.71	244,95	12.00
74	67.97	82.58	90.11	94.41	125.88	157.34	188.82	251.76	12.33
76	69.81	84.82	92.55	96.97	129.29	161.60	193.93	258.58	12.67
78	71.65	87.05	94.99	99.52	132.70	165.86	199.04	265.39	13.00
80	73.48	89.28	97.41	102.06	136.08	170.09	204.13	272.17	13.33
82	75.32	91.51	99.85	104.62	139.49	174.35	209.24	278.99	13.67
84	77.16	93.75	102.29	107.18	142.90	178.61	214.35	285.80	14.00
86	78.99	95.97	104.72	109.72	146.29	182.85	219.43	292.58	14.33
88	80.83	98.21	107.16	112.27	149.70	187.11	224.55	299.39	14.67
90	82.67	100.44	109.60	114.83	153.10	191.36	229.66	306.21	15.00

Weight of Round Steel Ductwork Per Linear Foot
With 20% Allowance

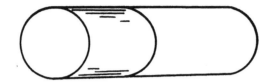

Diameter	Gauges, Lbs/SqFt and Lbs/SqFt with 20% Allowance									
	26	24	22	20	18	16	14	12	10	SOFT
Inches	0.75	1.00	1.26	1.60	2.00	2.50	3.13	4.37	5.63	LF
	0.90	1.20	1.50	1.80	2.40	3.00	3.76	5.24	6.76	
4	0.95	1.27	1.59	1.90	2.54	3.17	3.96	5.55	7.12	1.05
5	1.20	1.60	2.00	2.40	3.21	4.01	5.01	7.01	8.99	1.31
6	1.40	1.87	2.34	2.80	3.74	4.68	5.84	8.18	10.49	1.57
7	1.60	2.13	2.67	3.20	4.27	5.35	6.68	9.36	11.98	1.83
8	1.90	2.53	3.17	3.80	5.07	6.35	7.93	11.10	14.23	2.09
9	2.10	2.80	3.51	4.20	5.61	7.02	8.76	12.27	15.73	2.36
10	2.40	3.20	4.01	4.80	6.41	8.02	10.02	14.03	17.98	2.62
11	2.60	3.47	4.34	5.20	6.94	8.69	10.85	15.19	19.47	2.88
12	2.80	3.74	4.68	5.60	7.48	9.35	11.68	16.36	20.97	3.14
14	3.30	4.40	5.51	6.60	8.81	11.03	13.77	19.29	24.72	3.67
16	3.80	5.07	6.35	7.60	10.15	12.70	15.86	22.21	28.46	4.19
18	4.20	5.60	7.01	8.40	11.22	14.03	17.53	24.54	31.46	4.71
20	4.70	6.27	7.85	9.40	12.55	115.70	19.61	27.47	35.20	5.24
22	5.20	6.94	8.68	10.40	13.89	17.37	21.70	30.39	38.95	5.76
24	5.70	7.60	9.52	11.40	15.22	19.04	23.79	33.31	42.69	6.28
30	7.10	9.47	11.86	14.20	18.96	23.72	29.63	41.49	53.18	7.86
36	8.50	11.34	14.20	17.00	22.70	28.40	35.47	49.67	63.67	9.43
42	9.90	13.21	16.53	19.80	26.44	33.08	41.31	57.86	74.15	11.00
48	11.30	15.07	18.87	22.60	30.18	37.75	47.15	66.04	84.64	12.57

Weight of Aluminum Ductwork Per Linear Foot
With 20% Allowance

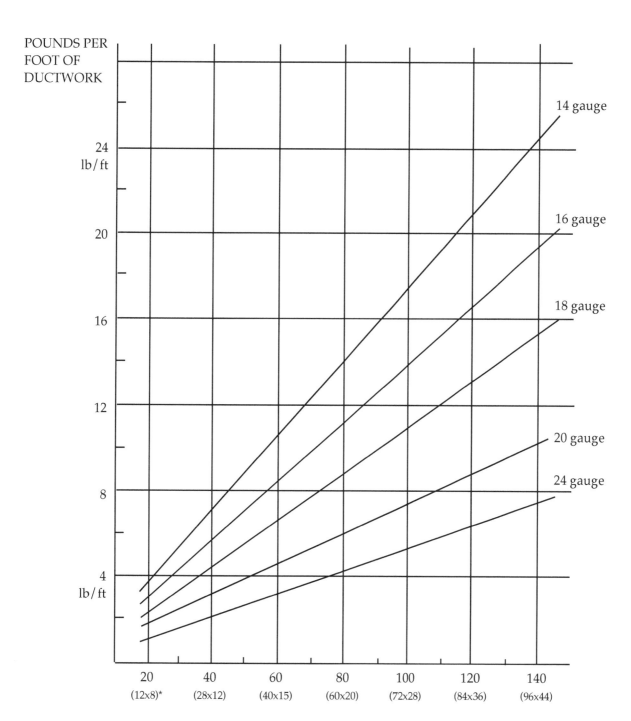

POUNDS PER FOOT OF DUCTWORK

24 lb/ft

20

16

12

8

4 lb/ft

14 gauge

16 gauge

18 gauge

20 gauge

24 gauge

| 20 | 40 | 60 | 80 | 100 | 120 | 140 |
| (12x8)* | (28x12) | (40x15) | (60x20) | (72x28) | (84x36) | (96x44) |

SEMI-PERIMETER OF DUCT, INCHES
(Width plus depth)

*Sample size

Material Data

Gauge	Galvanized			HRS/CRS		Stainless Steel	
	LBS/SF	w/20% Allow.	Thick-ness	LBS/S	Thick-ness	LBS/SF	Thick-ness
26	.906	1.1	.022	.750	.018	.748	.0178
24	1.156	1.4	.028	1.000	.024	.987	.0235
22	1.406	1.7	.334	1.250	.030	1.231	.0293
20	1.656	2.0	.040	1.500	-036	1.491	.0355
18	2.156	2.6	.052	2.000	.048	2.016	.0480
16	2.656	3.2	.063	2.500	.060	2.499	.0595
14	3.281	4.0	.078	3.125	.075	3.154	.0751
12	4.531	5.5	.108	4.375	.105	4.427	.1054
10	5.781	7.0	.138	5.625	.134	5.670	.1350
8				6.875	.164	6.930	.1650
3/16				7.650	.187		
1/4				10.200	.250		

Aluminum Weight Per Sq Ft

Alum. Gauge	Equiv. Galv. Gauge	LBS/SF	Thick-ness
24	26	.282	.020
22	24	.353	.025
20	22	.452	.032
18	20	.564	.040
16	18	.706	.050
14	16	.889	.063
12	14	1.13	.080
10	12	1.76	.125

Angle Weights Per Ft

Size	Galv.	HRS	Alum.
1x1x1/8	.84	.08	.28
1-1/4x1-1/4x1/8	1.06	1.01	.34
1-1/4x1-1/4x3/16	1.55	1.48	.51
1-1/4x1-1/4x1/4	2.02	1.92	.66
1-1/2x1-1/2x1/8	1.29	1.23	.42
1-1/2x1-1/2x3/16	1.89	1.80	.62
1-1/2x1-1/2x1/4	2.46	2.34	.81
2x2x1/8	1.73	1.65	.58
2x2x3/16	2.56	2.44	.85
2x2x1/4	3.35	3.19	1.11
2-1/2x2-1/2x3/16		3.07	1.07
2-1/2x2-1/2x1/4	4.31	4.10	1.40

CORRECTION FACTORS

Correction factors save time, increase the accuracy of bidding, reduce the bulk of bidding data needed, simplify the whole estimating process and point the way for understanding and controlling costs.

Correction factors are multipliers used to adjust labor and costs of common standard items and conditions. Correction factors show the percentage increase or decrease needed.

Instead of keeping complex bulky data on an infinite number of items and variable conditions, only a few are kept, while other similar items and different conditions are simply related with the multipliers.

The most common standard in the sheet metal industry, the bench mark, is low pressure galvanized ductwork, 24 gauge average, 25% fittings by weight, unlined, installed on lower floors at about 10 foot high in reasonably clear ceiling spaces. Most other ductwork and conditions are related to this situation.

For example, if it takes 10 hours to install the following low pressure, bare galvanized duct run under standard conditions, the following labor adjustments would have to be made for other conditions and hence more money be included in the bid.

	Labor 10 hrs x Correction Factor =	Adjusted Hours
10th Floor	1.16	11.6
25 ft High Duct	1.30	13.0
Congested Ceiling Space	1.15	11.5
Area 300 ft. from Unloading	1.05	10.5
35°C days	1.15	11.5
Existing Office Buildings	1.35	13.5
All Piping Installed	1.12	12.5

If the sell price (material, labor, overhead and profits) for the above standard low pressure galvanized duct run is $ 1000, the sell price for other type ducts would be:

	$1000 x Correction Factor	Adjusted Price
Lined 1" 1-1/2 lb	1.4	$1,400
Insulated 1-1/2", 3/4 lb	1.4	$1,400
Fiberglass Ductboard	.9	900
Automatic Duct Coil	.8	800
High Pressure with Cleats	1.35	1,350

Field Labor Correction Factors
for HVAC Equipment, Ductwork and Piping
Use as Multipliers Against Labor Hours

Floors

Bsmt, Ground, 1st	1.00
2nd & 3rd	1.04
4th, 5th, 6th	1.08
7th, 8th, 9th	1.12
10th, 11th, 12th	1.16
13th, 14th, 15th	1.21
16th, 17th, 18th	1.26
19th, 20th, 21st	1.31

Duct Heights

10 ft	1.00
15 ft	1.10
20 ft	1.20
25 ft	1.30
30 ft	1.40
35 ft	1.50
40 ft	1.60

Correction for Size of Job

0-24 hours	1.12
2548 hours	1.08
48-96 hours	1.04
96 hours and up	none

Special Areas

Open areas, no partitions	.85
Congested ceiling space	1.15
Equipment room	1.20
Kitchen	1.10
Auditorium; pool w/slopped floor	1.25
Attic space	1.50
Attic space	1.50
Crawl space	1.20
Cramped shaft	1.30
One or two continuous risers	.80

Distance from Unloading Point

100 ft	1.00
200 ft	1.03
300 ft	1.05

Existing Buildings

Typical existing office bldg, hospital, school, etc.	1.35
Existing factory, warehouse, gym, hall, garage, no ceilings	1.15
100% gutted area	1.10
Work around, over, machinery, furnit.	1.10
Protect machinery, floor furniture	1.05
Quiet Job	1.05
Occupied areas	1.05
Remove items and reinstall same	1.50
Remove and replace equipment	2.00

Cost Control

Piping or ductwork installed	1.10
Electrical conduit installed	1.05
Partitions & door frames erected	1.10
Ceiling grid installed	1.05
Overmanning job, crashing	1.20
Company overloaded with work	1.20
Not being on top of job	1.20
On top of job continually	.90
Go back to put something in	1.15
Out of phase work	1.15
Move stock pile about	1.05
Delays of deliveries, shop drawings, approvals, purch., fab., tools	1.15
Poor shop drawings	1.15
Poor foreman	1.20
Delay in facing prob. decision	1.15
Cluttered floors	1.15
Trades working on top of each other	1.10
No service roads, muddy	1.10
Congested traffic, poor unloading	1.10

Overtime

Overtime	40 to 50 hours	1.10
	50 to 60 hours	1.15
Night work	4 to 12 midnight	1.20
	12 to 8 am	1.20

Correction For Temperature

Under 20%	1.15
Over 90%	1.15

Ductwork Correction Factors
Multipliers Applied on Standard Low Pressure Galvanized Ductwork For Material, Connection, Automated Fabrication Variations

	MULTIPLIER ON: Total Price Furnished and Installed	FACTORS ON: Shop Labor	Field Labor
Low Pressure Rectangular Galvanized	1.00	1.00	1.00
Lined Low Pressure: 1" thick, 1-1/2 lb	1.40	—	1.03
1/2" thick, 1-1/2 lb	1.34	—	1.01
Insulated Duct: 1" thick, 1-1/2 lb	1.35	—	1.00
1-1/2" thick, 1-1/2 1	1.40	—	1.00
Medium Pressure: Cleat Connections	1.25	1.25	1.25
Angle Flanges	1.30	1.30	1.30
High Pressure: Cleat Connections	1.35	1.35	1.35
Angle Flanges	1.40	1.40	1.40
Coil Line Ductwork, Bare (shop factor, straight)			
Shop Assembled	.80	.30	1.00
Field Assembled	.80	.25	1.05
Plasma Arc Fitting Cutting Machine	.60	.50	1.00
Computerized Blank Out Program	.98	.95	1.00
Spiral: Low Pressure	.70	—	.80
High Pressure	.80	—	.90
Fab Only, low & high pressure	—	.25	—
Pipe Fittings	—	.80	—
Fiberglass Ductboard	.90	.70	.70
Aluminum, Light Gauges, Cleats	1.20	1.20	1.20
Stainless Steel, Light Gauges, Cleats	2.00	1.25	1.25
PVS, Light Gauges, Cleats	1.70	1.25	1.30
Angle Flange Connections versus Cleats	—	1.60	1.10
Bent Metal Flanges versus Cleats	—	1.10	1.05
Welded Connections versus: Cleats	—	.75	1.05
Ductmate versus: Cleats, (shop installed)	—	1.05	1.03
Angle Flanges (shop installed)	—	.75	.94
TDC Connections versus: Cleats	—	1.08	1.05
Angle Flanges	—	.68	.95
Cementing Seams and Connections	—	1.05	1.10
Taping Seams and Connections	—	1.05	1.05
Duplications: of Duct Sections, Shop	—	.66	—
of Field Areas	—	—	.80
Connect Two Duct Sections on Floor in Field	—	—	.75

This page intentionally left blank

Chapter 10

Galvanized Duckwork

This section of the manual covers pricing of low, medium and high pressure rectangular galvanized ductwork as used in HVAC systems in commercial, institutional and industrial buildings.

1. Low pressure galvanized ductwork comprises the bulk of HVAC ductwork used in buildings. It is used for system pressures between 0-2" S.P. and air velocities between 0-2500 FPM. Generally, connections are with cleats and the seams are snaplock or pittsburgh. Reinforcing is either crossbreaking, beading, reinforced cleats or structural angles.

2. Medium pressure galvanized ductwork is used for pressures from 2-6" S.P. and velocities from 2000 to 4000 within the S.P. range. High pressure galvanized ductwork is used in systems where the S.P. is over 6" and the velocities are over 2000 FPM.

    Both medium and high pressure ductwork must be sealed to maintain pressures within 1 or 1/2% of design CFM. Both are constructed with pittsburgh seams and the connections are with cleats which you can seal, or are gasket companion angles. Reinforcing is with angles, either backup near the connection and/or at prescribed intervals.

ASHRAE/SMACNA Standards

All ductwork weights, labor productivity rates and construction is based on SMACNA and ASHRAE criteria and on the traditional gauge breakdown. For example, low pressure galvanized 0-12" wide 26 gauge, 13"-30" 24 gauge, 31"-54" 22 gauge. 5 5"-84" 20 gauge and over 85" wide 18 gauge.

Methods of Estimating

This chapter covers estimating rectangular galvanized for HVAC systems both by the piece and by the pound. To estimate per linear foot, all per pound figures can be converted into linear foot rates.

All labor productivity rates and figures are based on standard conditions, lower floors, ten foot high duct runs, new construction and average space conditions.

ESTIMATING GALVANIZED
DUCTWORK BY THE PIECE

Benefits

1. The major benefit of estimating galvanized ductwork by the piece is supreme labor accuracy. You use the actual labor time normally expended for that specific type duct and for that size.

    Sheet metal contractors are sometimes off a plus or minus 25% in attempting to use the cost per pound method because of the variety of gauge mixtures, the different pipe and fitting combinations, which both radically effect the cost per pound and causes it to go up and down dependent on many variable factors.

    Where as labor per piece has the most valid correlation between the unit of measurement and the actual labor. It is the most direct, logical connector between a particular duct and its intrinsic labor. It automatically compensates for all mixtures of fittings and pipe, all duct sizes and gauges, for the various types of fittings. The inherent fallacy and confusion and uncertainty in per pound pricing is sidestepped. You don't care if there are 10% or 45% fittings by weight or if the average gauge is 24-1/2 or 22 1/4. You can ignore this criteria, and avoid possible erroneous guesses on labor.

    Pricing by the piece keeps your fingers on the real labor pulse automatically. If the end price should sell for 3.95/lb that is what you will end up with, if it should be $2.50, again that is what you are programmed to get.

2. Another advantage of per piece estimating is that it promotes clarity, simplicity, better understanding of duct runs and of the different pieces involved. The per piece method is a clear, simple counting of things, as one would grilles or dampers. It deals with quantities of things and understandable sizes and labor entities.

Per piece takeoff and extension dears the estimators mind and permits him to concentrate on including all items in a bid and on pricing them correctly. He feels more aware of what the job entails, offsets, risers, drops, complicated ducts, the function of the duct run, etc. It cultivates confidence and skill in him.

The per piece labor approach is something your sheet metal men in both the shop and field, your supervisors, foremen, production schedulers can relate to. It is an understandable, usable, practical, concrete unit of measurement. You can talk "pieces" with people, the hours per piece, you can count them easily, and you can relate sizes to labor accurately. This is very difficult to do, however, with pounds per hour estimating.

Per piece estimating will help you get the jobs you should get and keep you from getting those you should not!

BASIS OF PER PIECE LABOR AND MATERIAL

What's Included and Not
1.   Per piece labor is based on the actual labor required to fabricate or install a certain type duct in a certain size range, by an average sheet metal mechanic journeyman; what he should be able to do under normal, typical conditions with standard tools and erection equipment, under normal supervision. It is the mix of the fast and the slow worker, the competent and the less skillful man, the well run job and the poorly run one.

2.   Labor includes all production operations:
     a.   Shop labor includes unloading raw materials, sheets, listing, blanking, layout and cutting, seaming, forming, assembling, reinforcing angles, cleats, hangers and the final loading of the assembled items on the truck for shipment to the job site.
     b.   Field labor includes all operations from the tailgate of the truck to final cleanup, distribution of the ductwork, set up of scaffolding, tools, layout of the duct runs, cutting hangers and cleats, the actual hanging and finally the tear down of erection equipment and clean up.
     c.   Both material handling and supervision are also included in the shop and field labor figures in per piece pricing.

3.   Non-production operations such as drafting, truck driving, and field measuring are not included in the per piece unit hours.

4.   Installation unit hours are based on the time it takes to erect complete duct runs as a batch and not as single pieces.

5.   The material for reinforcing angles are not included in the unit weights and must be calculated separately, but the labor time, as stated above, to fabricate and attach them to the ducts is included.

6.   Per piece labor and material is based on SMACNA specifications for gauges and construction.

TAKEOFF AND EXTENSION PROCEDURES FOR ESTIMATING DUCTWORK BY THE PIECE

A.   Takeoffs
1.   Use the Per Piece Duct Takeoff Sheet for listing sizes, lengths, wt per ft, shop labor, field labor, etc.

2.   List each duct size, as you do in per pound estimating, indicate the type duct, write in the equivalent length for fittings and the measured length for pipe for each piece required.

B.   Extension for Per Piece Estimating
1.   The material weight is calculated by adding up the lineal footages on each line, inserting the lb/lf from the lb/lf chart, which has a 20% allowance included in it already, and then the two are multiplied and the lbs are written in the weight column.

2.   Labor is determined by adding up the number of pieces on each line and inserting the total in the "qty" of pieces column. The unit labor per piece for the various types of ducts and size categories is taken from the galvanized labor charts and written in the unit labor columns for the shop and field and multiplied time the quantity of pieces for the total labor for both the shop and the field.

3.   After all labor and material are extended the columns are added up for the grand totals of each and transferred to the summary sheet.

C.   Calculate reinforcing angles as needed.

D.   Calculate the turning vanes and splitter dampers as needed.

E.   Check work by measuring total linear feet of ducts on drawings with measuring wheel and compare with total linear feet from takeoff sheets.

Per Piece Takeoffs and Equivalent Lengths

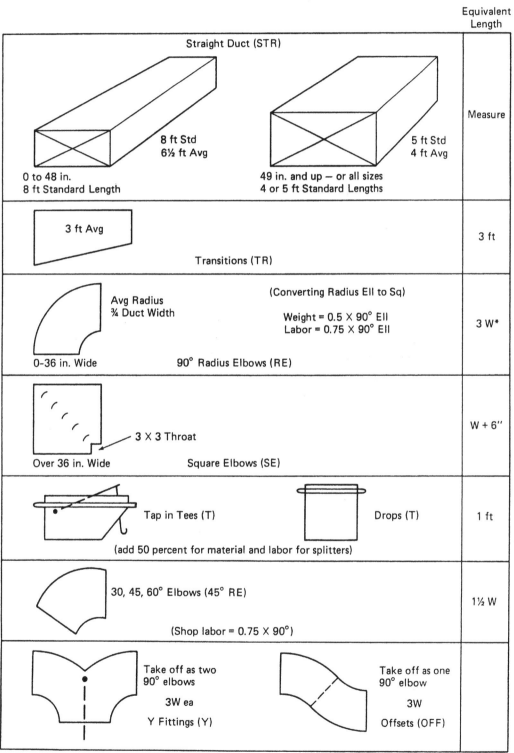

Equivalent Length

Straight Duct (STR)

8 ft Std
6½ ft Avg

5 ft Std
4 ft Avg

0 to 48 in.
8 ft Standard Length

49 in. and up — or all sizes
4 or 5 ft Standard Lengths

Measure

3 ft Avg

Transitions (TR)

3 ft

Avg Radius
¾ Duct Width

(Converting Radius Ell to Sq)

Weight = 0.5 × 90° Ell
Labor = 0.75 × 90° Ell

3 W*

0-36 in. Wide   **90° Radius Elbows (RE)**

3 × 3 Throat

W + 6"

Over 36 in. Wide   **Square Elbows (SE)**

Tap in Tees (T)   Drops (T)

1 ft

(add 50 percent for material and labor for splitters)

30, 45, 60° Elbows (45° RE)

(Shop labor = 0.75 × 90°)

1½ W

Take off as two
90° elbows

3W ea

Y Fittings (Y)

Take off as one
90° elbow

3W

Offsets (OFF)

*W: width of duct

Example Per Piece Takeoff and Extension
Listing Duct Sizes and Lengths

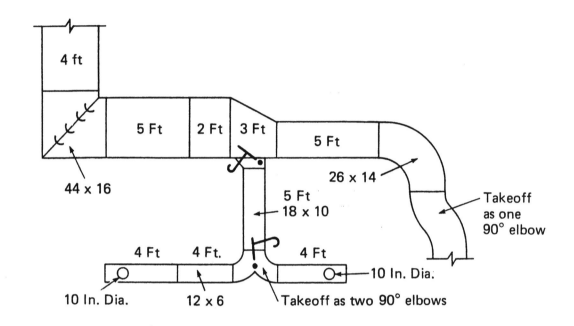

Per Piece Duct Takeoff Sheet

Job __1st NATIONAL BANK__   Mat'l __Galv.__   Duct Elev. __10'__

Drwg/Flr __M-1__   Sys. __S-1__   Pres __LOW__   Lin/Insul __1"__

QTY	DUCT SIZE	TYPE DUCT	Equivalent LINEAL FEET Per Piece	TOT LF	WEIGHT LBS /LF	WEIGHT Total	SHOP Hrs /Pc	SHOP Total	FIELD Hrs /Pc	FIELD Total
3	44x16	STR	4-5-2	11	17	187	.8	2.4	2.3	6.9
1	"	SE	4	4	17	68	2.8	2.8	3.3	3.3
1	"	TR	3	3	17	51	2.2	2.2	2.2	2.2
1	26x14	STR	5	5	9.4	47	.55	.55	1.6	1.6
1	"	RE	6	6	9.4	56	2.4	2.4	2.3	2.2
1	"	RE	6 (offset)	6	9.4	57	2.4	2.4	2.2	2.2
1	18x10	TEE	1	1	6.6	7	.6	.6	.6	.6
1	"	STR	5	5	6.6	33	.3	.3	.9	.9
2	12x6	RE	3-3 ("Y"Fitt.)	6	3.3	19	.8	1.6	.6	1.2
3	"	STR	4-4-4	12	3.3	40	.2	.4	.5	1.0
2	10"ϕ	TEE	1-1	2	3.3	7	.4	.8	.5	1.0
17	pcs			61'		572 LBS		16.4 HRS		23.1 HRS

Rectangular Galvanized, Low Pressure Ductwork
Shop Hours
Fittings Hours Per Piece; Straight Hours Per Foot

Manual Fabrication

Typical Size	Stretch Out	Straight	90 Elbw Offsets	45 Elbw	Sq/Elbw	Trans	Tap	Wye	Sq/Rd
12x6	3	0.04	0.85	0.64	0.65	0.58	0.43	1.70	1.46
15x6	3.5	0.05	0.97	0.73	0.74	0.67	0.46	1.94	1.67
18x6	4	0.05	1.09	0.82	0.82	0.75	0.50	2.17	1.88
18x12	5	0.07	1.32	0.99	0.99	0.92	0.56	2.64	2.30
24x12	6	0.08	1.55	1.17	1.16	1.09	0.63	3.11	2.72
30x12	7	0.09	2.30	1.73	1.72	1.26	0.80	4.61	3.14
36x12	8	0.11	2.45	1.84	1.80	1.42	0.83	4.91	3.56
42x12	9	0.15	2.60	1.95	2.96	2.08	1.32	5.21	5.19
42x18	10	0.16	2.75	2.07	3.01	2.24	1.35	5.51	5.60
48x18	11	0.17	2.90	2.18	3.06	2.40	1.37	5.81	6.00
60x18	13	0.19	4.11	3.09	3.16	2.73	1.58	8.23	6.81
72x18	15	0.22	4.95	3.71	4.45	3.24	2.28	9.89	8.10
84x24	18	.26	5.83	4.37	4.60	3.82	2.35	11.66	9.65
96x36	22	.38	7.91	5.93	7.56	5.71	2.45	15.82	14.28
120x36	26	0.45	8.87	6 65	7.94	6.70	2.54	17.74	16.74

Automated Fabrication

Typical Size	Stretch Out	Straight	90 Elbw Offsets	45 Elbw	Sq/Elbw	Trans	Tap	Wye	Sq/Rd
12x6	3	0.01	0.43	0.32	0.33	0.29	0.22	0.85	0.73
15x6	3.5	0.02	0.49	0.37	0.37	0.34	0.23	0.97	0.84
18x6	4	0.02	0.55	0.41	0.41	0.38	0.25	1.09	0.94
18x12	5	0.02	0.66	0.50	0.50	0.46	0.28	1.32	1.15
24x12	6	0.02	0.78	0.59	0.58	0.55	0.32	1.56	1.36
30x12	7	0.03	1.15	0.87	0.86	0.63	0.40	2.31	1.57
36x12	8	0.03	1.23	0.92	0.90	0.71	0.42	2.46	1.78
42x12	9	0.05	1.30	0.98	1.48	1.04	0.66	2.61	2.60
42x18	10	0.05	1.38	1*04	1.51	1.12	0.68	2.76	2.80
48x18	11	0.05	1.45	1.09	1.53	1.20	0.69	2.91	3.00
60x18	13	0.06	2.06	1.55	1.58	1.37	0.79	4.12	3.41
72x18	15	0.07	2.48	1.86	2.23	1.62	1.14	4.95	4.05
84x24	18	0.08	2.92	2.19	2.30	1.91	1.18	5.83	4.77
96x36	22	0.11	3.96	2.97	3.78	2.86	1.23	7.91	7.14
120x36	26	0.14	4.44	3.33-	3.97	3.35	1.27	8.87	8.37

Rectangular Galvanized, Low Pressure Ductwork
Field Hours
Fittings Hours Per Piece; Straight Hours Per Foot

Typical Size	Stretch Out	Straight	Elbow Offsets	Trans Sq/Rd	Tap	Wye
12x6	3	0.08	0.66	0.53	0.46	1.32
15x6	3.5	0.11	0.82	0.65	0.57	1.64
18x6	4	0.14	0.99	0.76	0.69	1.98
18x12	5	0.19	1.32	0.99	0.93	2.64
24x12	6	0.24	1.65	1.22	1.16	3.30
30x12	7	0.29	1.98	1.44	1.39	3.96
36x12	8	0.34	2.31	1.67	1.63	4.62
42x12	9	0.39	2.64	1.90	1.86	5.28
42x18	10	0.45	2.97	2.13	2.10	5.94
48x18	11	0.50	3.30	2.36	2.33	6.60
60x18	13	0.60	3.96	2.81	2.80	7.92
72x18	15	0.70	4.62	3.27	3.27	9.24
84x24	18	0.86	5.61	3.95	3.97	11.22
96x36	22	1.32	8.71	6.07	6.07	17.42
120x36	26	1.54	10.20	7.03	7.03	20.40

Rectangular Galvanized, Medium Pressure Ductwork
Shop Hours
Fittings Hours Per Piece; Straight Hours Per Foot

Manual Fabrication

Typical Size	Stretch Out	Straight	90 Elbw Offsets	45 Elbow	Sq/Elbow	Trans	Tap	Wye	Sq/Rd
124	3	0.05	1.02	0.77	0.78	0.70	0.52	2.04	1.75
154	3.5	0.06	1.16	0.88	0.89	0.80	0.55	2.33	2.00
18x6	4	0.06	1.31	0.98	0.98	0.90	0.60	2.60	2.26
18x12	5	0.08	1.58	1.19	1.19	1.10	0.67	3.17	2.76
24x12	6	0.10	1.86	1.40	1.39	1.31	0.76	3.73	3.26
30x12	7	0.11	2.76	2.08	2.06	1.51	0.96	5.53	3.77
36x12	8	0.13	2.94	2.21	2.16	1.70	1.00	5.89	4.27
42x12	9	0.18	3.12	2.34	3.55	12.50	1.58	6.25	6.23
42x18	10	0.19	3.30	2.48	3.61	2.69	1.62	6.61	6.72
48x18	11	0.20	3.48	2.62	3.67	2.88	1.64	6.97	7.26
60x18	13	0.23	4.93	3.71	3.79	3.28	1.90	9.88	8.17
72x18	15	0.26	5.94	4.45	5.34	3.89	2.74	11.87	9.72
84x24	18	0.31	7.00	5.24	5.52	4.58	2.82	13. 99	11.45
96x36	22	0.46	9.49	7.12	9.07	6.85	2.94	18.98	17.14
120x36	26	0.54	10.64	7.98	9.53	8.04	3.05	21.29	20.09

(Continued)

Automated Fabrication

Typical Size	Stretch Out	Straight	90 Elbw Offsets	45 Elbw	Sq/Elbw	Trans	Tap	Wye	Sq/Rd
12x6	3	0.01	0.51	0.38	0.39	=.35	0.26	1.02	0.88
15x6	3.5	0.02	0.58	0.44	0.44	0.40	0.28	1.16	1.00
18x6	4	0.02	0.65	0.49	0.49	0.45	0.30	1.30	1.13
18x12	5	0.03	0.79	0.59	0.59	0.55	0.34	1.58	1.38
24x12	6	0.03	0.93	0.70	0.70	0.65	0.38	1.87	1.63
30x12	7	0.03	1.38	1.04	1.03	0.76	0.48	2.77	1.88
36x12	8	0.04	1.47	1.10	1.08	0.85	0.50	2.95	2.14
42x12	9	0.05	1.56	1.17	1.78	1.25	0.79	3.13	3.11
42x18	10	0.06	1.65	1.24	1.81	1.34	0.81	3.31	3.36
48x18	11	0.06	1.74	1.31	1.84	1.44	0.82	3.49	3.60
60x18	13	0.07	2.47	1.85	1.90	1.64	0.95	4.94	4.09
72x18	15	0.08	2.97	2.23	2.67	1.94	1.37	5.93	4.86
84x24	18	0.09	3.50	2.62	2.76	2.29	1-41	7.00	5.72
96x36	22	0.14	4.75	3.56	4.54	3.43	1.47	9.49	8.57
120x36	26	0.16	5.32	3.99	4.76	4.02	1.52	10.64	10.04

Rectangular Galvanized, Medium Pressure Ductwork
Field Hours
Fittings Hours Per Piece; Straight Hours Per Foot

Typical Size	Stretch Out	Straight	Elbow Offsets	Trans Sq/Rd	Tap	Wye
12x6	3	0.10	0.79	0.64	0.55	1.58
15x6	3.5	0.13	0.98	0.78	0.68	1.97
18x6	4	0.17	1.19	0.91	0.83	2.38
18x12	5	0.23	1.58	1.19	1.12	3.17
24x12	6	0.29	1.98	1.46	1.39	3.96
30x12	7	0.35	2.38	1.73	1.67	4.75
36x12	8	0.41	2.77	2.00	1.96	5.54
42x12	9	0.47	3.17	2.28	2.23	6.34
42x18	10	0.54	3.56	2.56	2.52	7.13
48x18	11	0.60	3.96	2.83	2.80	7.92
60x18	13	0.72	4.75	3.37	3.36	9.50
72x18	15	0.84	5.54	3.92	3.92	11.09
84x24	18	1.03	6.73	4.74	4.76	13.46
96x36	22	1.58	10.45	7.28	7.28	20.90
120x36	26	1.85	12.24	8.44	8.44	24.48

USING MULTIPLIERS FOR DUCTWORK

This chapter contains correction factors for using a plasma cutter for galvanized ductwork versus manual fabrication, x 0.50.

This factor is applied to the total labor to fabricate galvanized ductwork. For per piece labor use this factor as a multiplier. For example, if it takes a total of 2 hours to fabricate a 24x12 elbow by hand, it only takes 0.50 times 2 which equals 0.50 hours with a plasma cutter.

If you are using lbs per hour use 0.50 as a divisor. For example, 4 lbs per hr divided by 0.50 equals 88 lbs per hr. If you are using hours per lb units, then.024 hrs per lb divided by 0. 50 equals .0 12 hrs per lb.

ESTIMATING GALVANIZED DUCTWORK BY THE POUND

Galvanized ductwork on a particular project can run anywhere from $1.50 to $4.50 per pound installed depending on the mixture of straight duct and fittings and on the average size.

Yet too many contractors use the ground beef approach for estimating galvanized ductwork by the pound. They throw all different grades of beef into the meat grinder whether $2.00 or $4.00/lb, crank it through, come up with one big pile and have no real idea of what the correct mixed price per pound is.

When ductwork is all lumped together into one heap contractors have great difficulties determining what the correct labor productivity rates are on a per pound basis. Hence he bids roughly the same price per pound regardless of fitting ratio or the average duct size, or makes a wild guess, and then either loses money or doesn't get the job.

Methods of Estimating Galvanized Ductwork

There are three basic approaches to estimating galvanized ductwork by the pound to resolve this problem.

1. *The traditional lump method.* In this approach straight and fittings are taken off together and everything is lumped together into one weight as described above. Then a judgment is made as to what the percentage fittings and average size are for the project, and a combined labor productivity rate is selected from the chart on page 140.

    This approach can work reasonably well if you have actual cost records from similar installations, or if it is obviously typical standard ductwork. But if it's not, and costs should be 20% more or less, a great risk is taken in guessing.

2. *Labor based on percentage fittings and average size.* In this approach fittings and straight are taken off separately, the percentage fittings calculated, the weights for each gauge totaled, and a judgment made by inspection as tot he average gauge. Then everything is lumped into one weight and a combined productivity labor rate is taken from the chart, based on the percentage fittings and average size.

3. *Separate labor productivity rates for each gauge of fillings and of straight.* In this method the fittings and straight are taken off separately and the weights per gauge are kept separate. Then separate labor productivity rates for each gauge for fittings and for straight are applied.

    This approach is the most accurate, but it is cumbersome and time consuming, if done manually. It lends itself well to computer operations.

Rectangular Galvanized Low Pressure Ductwork
Labor Per Pound

Fabrication Labor

										Duct Width Range
	0-12"		13-30"		31-54"		55-84"		85" up	
Percentage Fittings By Weight	26 ga LB /HR	HR /LB	24 ga LB /HR	HR /LB	22 ga LB /HR	HR /LB	20 ga LB /HR	HR /LB	18 ga LB /HR	HR /LB
Str duct only	76	.013	90	.011	100	.010	105	.0095	120	.0083
10 - 20%	48	.021	59	.017	67	.015	72	.014	77	.013
20 - 30%	38	.026	44*	.023	53	.019	56	.018	63	.016
30 - 40%	33	.030	37	.027	48	.021	53	.019	56	.018
40 - 50%	28	.036	33	.030	42	.024	48	.021	50	.020
Fittings only	15	.067	20	.050	24	.042	28	.035	30	.033
	.91		1.165		1.41		1.66		2.16 LB/SF	

Installation Labor

	26 ga LB /HR	HR /LB	24 ga LB /HR	HR /LB	22 ga LB /HR	HR /LB	20 ga LB /HR	HR /LB	18 ga LB /HR	HR /LB
Str duct only	26	.039	29	.034	32	.031	34	.029	36	.028
10 - 20%	23	.043	27	.037	31	.032	33	.030	34	.029
20 - 30%	21	.048	25*	.040	30	.033	32	.031	33	.030
30 - 40%	20	.050	24	042	28	.035	31	.032	32	.031
40 - 50%	19	.053	23	~043	27	.037	30	.033	32	.031
Fittings only	15	.067	19	.053	23	.043	27	.037	29	.034

*Average, typical project
Labor based on gross weight with 20 percent allowance included.

Measuring Ductwork for Pounds Per Hour Pricing
Equivalent Lengths

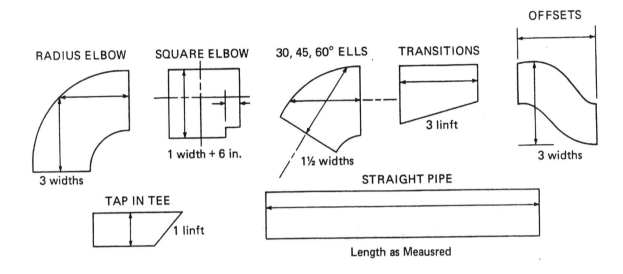

Pounds Per Hour Fabrication Labor
for Low Pressure Galvanized Ductwork

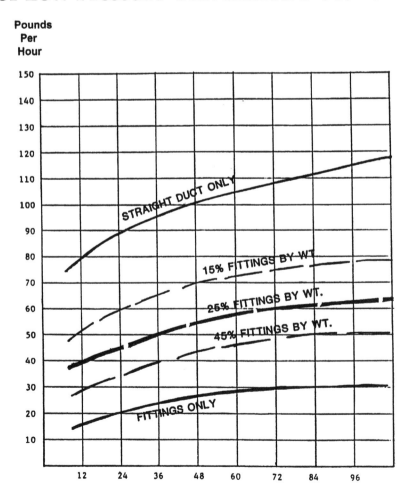

Duct Takeoff Sheet
Per Pound

Job **Elm High School**   Mat'l **G**   Duct Elev. **12'**
Drwg/Flr. **M-1**   Sys **S-1**   Pres. **LP**   Lin/Insul ———

Duct Size	SF /LF	LINEAL FEET Straight Duct	Total LF	LBS /LF w 20%	0-12 26ga	13-30 24ga	31-54 22ga	55-84 20ga	85·up 18ga	SF
18X12	11-15		26	7.02		182 LBS				
28X10	7-5-2		14	8.88		125				
28X20	12		12	11.2		135				
36X24	12		12	17.0			204			
50x34	10-18		28	23.8			666			
24x12	5-15-5		25	8.4		210				
12x12	10		10	4.4	44					
26x18	9-8		17	10.2		174				
20x10	26		26	7.02		182				
96x48	6		6	62.4					375	
72x36	20		20	35.3				706		
60x24	3		3	28.0				84		
		Str. Wt. Total	199		44	1008	870	790	375	= 3087 LBS
		Fittings								
18x12	3-5-4-6-2		20	7.02		140 LBS				
28x20	2-3-5		10	11.2		112				
50x34	5-4		9	23.8			214			
20x10	3-3-2		8	7.02		58				
72x36	6		6	35.3				212		
12x12	2-2-2-2		8	4.4	.35					
		Fitt. Wt. Total	61		.35	310	214	212		= 771 LBS

Total Job Weight — 3858 LBS

Percent Fittings by Weight | 771 ÷ 3858 = 20%

Average Gauge - 22

Fabrication Labor   3858 LBS ÷ 60* LBS/HR = 64.3 HRS

Installation Labor   3858 LBS ÷ 30* LBS/HR = 128.6 HRS

Duct Takeoff Sheet
Per Pound

Job __Elm High School__   Mat'l __G__   Duct Elev __12'__

Drwg/Flr __M-1__   Sys __S-1__   Pres __LP__   Lin/Insul ____

Duct Size	SF /LF	LINEAL FEET Straight Duct	Total LF	LBS /LF w 20%	0-12 26ga	13-30 24ga	31-54 22ga	55-84 20ga	85 up 18ga	SF
18x12		11-15	26	7.02		182	LBS			
28x10		7-5-2	14	8.88		125				
28x20		12	12	11.2		135				
36x24		12	12	17.0			204			
50x34		10-18	28	23.8			666			
24x12		5-15-5	25	8.4		210				
12x12		10	10	4.4	44					
26x18		9-8	17	10.28		174				
20x10		26	26	7.02		182				
96x48		6	6	62.4					375	
72x36		20	20	35.3				706		
60x24		3	3	28.0				84		
		Wt. Totals	199		44	1008	870	790	375 =3087 LBS	
		Shop Labor Str ÷			76	90	100	105	120 LBS/HR	
					.58	11.2	8.7	7.5	3.13 HRS	
						Total Shop =	31.1		HRS	
		Field Labor Str ÷			26	29	32	34	36 LBS/HR	
					1.7	34.6	27.2	23.3	10.4 HRS	
						Total Field =	97.2		HRS	
		Fittings								
18x12		3-5-4-6-2	20	7.02		140	LBS			
28x20		2-3-5	10	11.2		112				
50x34		5-4	9	23.8			214			
20x10		3-3-2	8	7.02		58				
72x36		6	6	35.3				212		
12x12		2-2-2-2	8	4.4	35					
		Wt. Totals	61'		35	310	214	212	= 771 LBS	
		Shop Labor Fitt. ÷			15	20	24	28	LBS/HR	
					2.4	15.5	8.9	7.6 =34.4 HRS		
		Field Labor Fitt. ÷			15	19	23	27	LBS/HR	
					2.4	16.3	9.3	7.9 =35.9 HRS		
		Totals	260'		Shop 65.5 HRS	Field 133 HRS	3858 LBS			

FACTORS AFFECTING GALVANIZED DUCTWORK LABOR WHEN ESTIMATING ON A PER POUND BASIS

Fabrication labor for a mixture of straight and fittings can vary from .015 to .030 hours per lb dependent on the percentage fittings. A typical 25 percent fittings by weight mixture normally runs about .023 hours per lb.

A deviation in the mixture of different fittings can vary up or down 50 percent from between .025 to .075 hour per lb.

The following conditions can change the hours per lb labor productivity rates up or down as much as 50 percent or more overall:

1. The pressure classification of ductwork, low, medium, high, etc.

2. The ratio of percentage weight of fittings to total weight of both pipe and fittings.

3. Average size of ductwork, 26, 24, 22 gauge, 12"x12", 24"x12", 48"x18" sizes, etc.

4. The actual mix of different fittings
   Typical mix in commercial low pressure HVAC

system:
50% tees
25% elbows
12% transitions
12% offsets, wyes, etc.

5. Average length of straight pipe

6. Average length of transitions

7. The floor the ductwork is being installed on

8. Height of ductwork

9. Type connection, cleats, flange, ductmate, TDC, etc.

10. Ceiling space conditions

Per piece laboring for galvanized ductwork automatically adjusts these hours per lb according to the average size, percentage, fittings, type fittings, average lengths, type connections, etc.

Installed Cost Per Pound For Galvanized Ductwork
Budget Pricing

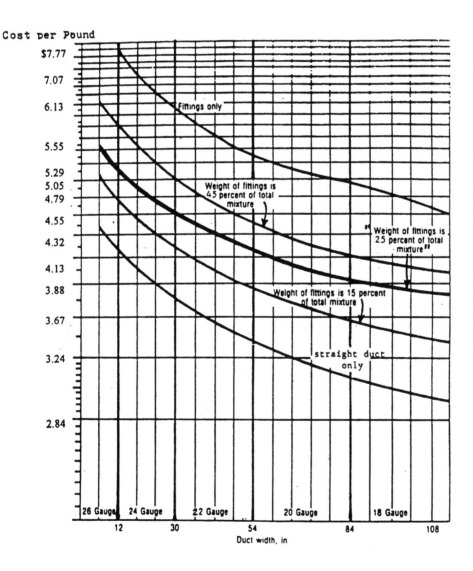

Curves for cost per pound of bare low pressure galvanized ductwork for new construction, 2000 lbs and up, standard installation conditions and conventional duct fabrication (not coil line).

Prices include material, shop and field labor, shop drawings, shipping and a 30 percent markup for overhead and profit on both material and labor.

Costs are based on galvanized material at $.40 per pound and direct labor wage rates of $39.00 per hour which includes base pay, fringes, insurance and payroll taxes.

The Ever Increasing Cost of Standard
Low Pressure Galvanized Ductwork Installed

**Cost Per Pound
Installed**

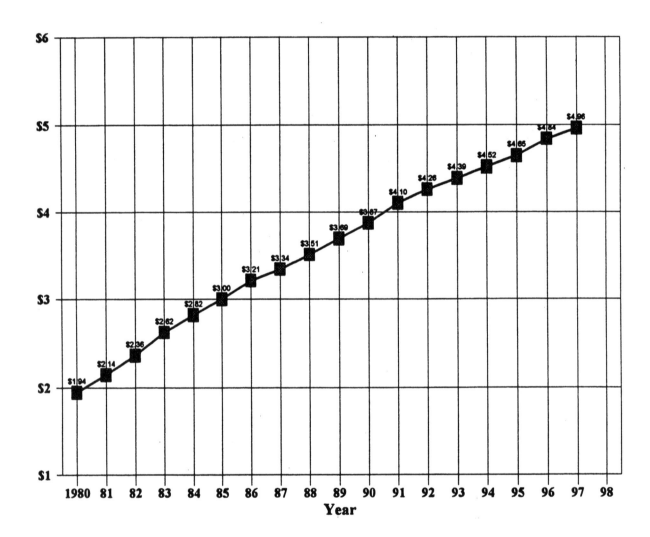

Year

MEDIUM AND HIGH PRESSURE DUCTWORK

Pressure Classification

Medium pressure   2000 fpm and up   2 to 6 inches S.P.
High pressure   2000 fpm and up   6 to 10 inches S.P.

Construction

About 2 gauges heavier than low pressure galvanized

More reinforcing angles required combined with alternate tie rods.

Angle reinforced double S cleats, welded flanges or companion angles are used at connections.

Seams and connections are sealed to withstand pressures 25% over design S.P.

Correction Factors

(On the totally installed cost of low pressure if figured as the identical duct system.)

Medium pressure cleat connections 1.25
Medium pressure companion angles 1.30
High pressure cleat connections.......................1.35
High pressure companion angles....................1.40

Take Off Procedure

Take off the same as low pressure ductwork, either by the piece or per pound.

Methods of Pricing

Completely price all labor and material as low pressure and apply a correction factor from above onto the total price.

Example:

As Low Pressure	As Medium Pressure Cleats
$30,000	1.25 x $30,000 = $37,500

Or apply factors to the low pressure material and labor separately:

As Low Pressure	As Medium Pressure Angles
1500 lbs	1.30 x 1500 = 1950 lbs
35 hrs shop	1.30 x 35 = 46 hrs
60 hrs field	1.30 x 60 = 78 hrs,

Budget Pricing

(Based on 10,000 lb job, 25% fittings by weight, standard conditions)

Medium pressure cleat connections $4.93
Medium pressure companion angles 5.11

High pressure cleat connections......................5.33
High pressure companion angles....................5.54

Galv. 0.41/lb, Labor $39.00/hr. Includes material, all labor, cartage, drafting, overhead and profit.

Included in the above labor figures are:

1. Switching from one gauge cod to another
2. Duct size changes
3. Replacing coils in cradle
4. Labor to remove stacks of L's on tables or skids from machine
5. Banding stacks of L's and loading on truck
6. Loading assembled duct sections on truck
7. Daily maintenance and breakdowns
8. Operator idle time
9. Supervision time
10. Machine running speeds of 25 to 30 fpm

Automatic Duct Coil Line Fabrication

Operations Performed by Coil Line

| Uncoil & Straighten | Shear | Notch | Bead | Seam | Drives | Bend |

Fabricated Labor

Gauge	Width Range of Duct	Straight Duct LBS/HR		Straight & Fittings Combined LBS/HR			
		Fab. L's Only	Fab L's & Assem in Shop	Percent Fittings by Weigh of Tot.			
				10-20%	20-30%	30-40%	40-50%
26	0-12	860	230	72	50	38	31
24	13-30	1100	300	95	66	51	41
22	31-54	1300	350	116	79	61	49
20	55-84	1500	420	135	93	71	58
18	85 up	1900	560	153	103	78	63

Correction Factors

    1. Assemble L's in field versus in the shop 1.07

Waste Allowance

    There is little or no waste in fabrication straight duct on a coil line. However, a 10% allowance must be added to the raw duct weight for corner seams, cleat edges and hangers.

Budget Figures

    Typical, commercial type ductwork, average mix of gauges and sizes, installed.

$3.96/lb   11 lbs/hr   9000 lbs/day   2-12 Mills lbs/yr

Chapter 11

Spiral and Light Gauge Round Duckwork

TYPES OF ROUND HVAC DUCTWORK

Spiral Pipe and Welded Fittings

Spiral pipe and welded fittings are primarily used for medium pressure HVAC systems but they are also used to some degree in general and industrial exhaust systems.

Connections are generally the slip type but companion angles can also be used. Slip connections are cemented or taped in high pressure or industrial exhaust systems. Heat shrink bands may also be used.

Spiral comes in round and oval configurations and can be either single or double skin.

Flexible Tubing

Flexible tubing for HVAC systems is constructed of helical wire or bands covered with different types of fabrics depending on usage. Special fabrics may be fire, moisture or corrosion resistant.

Flexible tubing may be single skin or factory wrapped with insulation.

Connections are the slip type and are fastened with draw bands.

Residential Furnace Pipe and Fittings

Residential furnace pipe is used for residential, apartment and light commercial HVAC work. It can also be used as a smoke flue.

Pipe is constructed with length wise snaplock seams and elbows are generally the adjustable type.

Residential furnace pipe is generally constructed of galvanized material but it is also available in stainless steel, aluminum and in a blue or walnut finish.

Connections are crimped on one end, slipped together and screwed or taped.

Flues

Flues are used for venting fumes from smaller HVAC heating equipment. It comes in either single or double skin and in round or oval.

Round Duct Gauge Data

RECOMMENDED GAUGES FOR GALVANIZED ROUND DUCTS

DUCT DIAMETER, INCHES	LOW PRESSURE DUCTS AND FITTINGS	MEDIUM AND HIGH PRESSURE DUCTS		
		SPIRAL PIPE	LONGITUDINAL SEAM DUCT	WELDED FITTINGS
Up thru 8"	26 Ga.	26 Ga.	24 Ga.	22 Ga.
9-13	26	24	22	20
14-22	24	24	22	20
23-36	22	22	20	20
37-50	20	20	20	18
51-60	18	18	18	18
61-84	16	—	16	16

RECOMMENDED GAUGES FOR OVAL GALVANIZED OR B.I. DUCT

DIMENSION OF MAJOR AXIS, INCHES	SPIRAL	LONGITUDINAL SEAM SUCT	WELDED FITTINGS
Up thru 24"	24 Ga.	20 Ga.	20 Ga.
25 to 36	22	20	20
37 to 48	22	18	18
49 to 50	20	18	18
51 to 70	20	16	16
71 and Over	18	16	16

Labor to Install Furnace Pipe, Flexible Tubing and Flues

HOURS PER PIECE, PIPE AND FITTINGS MIXED

AVG FURN. DIA.	FLEX. SF/LF	FLUE PIPE 5 ft	TUBING 5-6 ft	DOUB. WALL
4"	1.1	.4	.3	.7
6	1.6	.5	.4	.8
8	2.1	.6	.5	1.0
10	2.6	.8	.6	1.2
12	3.2	.9	.7	1.4
16	4.2	1.1	.9	—
15	4.7	1.2	1.0	—
22	5.8	1.5	1.2	—
24	6.3	—	—	—
30	7.9	—	—	—
36	9.4	—	—	—
42	11.0	—	—	—
48	12.5	—	—	—

(Labor based on ratio of 1 joint of pipe to 1 fitting)

Spiral Pipe and Fittings Correction Factors

Fabrication

Labor Correction Factors Applied to Manually Fabricated Spiral Fittings

Layout from templates	.70
Wire feed weld	.90
Stock cut, production run	80
Plasma cutter	45

Budget Figures to Purchase Only
Includes material, fabrication and overhead and profit.

1. .Spiral, Pipe & Fittings	$.98/lb
Pipe Only, 420 lb/hr (360 sq ft/hr)	$.63/lb
Fittings Only, 20.5 lb/hr (21.5 sq ft/hr)	$ 3.01/lb
Double Skin Fittings	$ 3.86/lb
2. .Residential Furnace Pipe	$.98/lb

Spiral Accessories
Four percent approximate, average size 18″ diameter.

Cement:	15 connections per gallon (105 ft of pipe/gal) at $17/gal
Tape:	25 connections per roll (175 ft duct per roll)
	2 inch wide, 180 ft/roll at $11/roll = 6.3¢ ft.
Shrink bands:	$1.98 per ft in rolls
Hangers:	Add 6%

Installation

Field Labor Correction Factors Applied to Medium Pressure Slip Connections

Spiral low pressure	0.90
Spiral double skin	1.75
Oval	1.5
Oval double skin	2.0
Underground	1.5
Thermal shrink bands on spiral	0.8

Low Pressure Spiral Ductwork
Shop Hours
Straight Hours Per Foot; Fittings Hours Per Piece

Manual Fabrication

Diameter	Straight	90 Elbw Offset Wye	45 Elbw Miters Reducer Tap	90T	Lat	90RdT	Sq/Rnd	Cross RdLat LatCross
4″	0.007	1.29	0.67	1.42	1.51	2.25	2.38	2.55
5″	0.007	1.46	0.78	1.45	1.54	2.30	2.47	2.55
6″	0.007	1.67	0.86	1.47	1.56	2.35	2.55	2.55
7″	0.153	1.81	0.95	1.50	1.59	2.39	2.64	2.55
8″	0.008	2.05	1.05	1.53	1.62	2.43	2.72	2.55
9″	0.009	2.24	1.12	1.56	1.75	2.47	2.81	2.72
10″	0.009	2.42	1.23	1.59	1.81	2.52	2.89	2.83
12″	0.010	2.81	1.42	1.64	1.99	2.86	3.09	3.15
14″	0.012	3.18	1.61	1.92	2.17	3.14	3.30	3.45
16″	0.013	3.56.	1.79	2.10	2.35	3.41	3.50	3.77
18″	0.014	3.72	2.16	2.28	2.53	3.68	3.72	4.09
20″	0.016	4.75	2.38	2.46	2.71	3.95	3.91	4.39
22″	0.018	5.80	2.69	2.59	2.97	4.22	4.17	4.75
24″	0.020	6.16	3.06	2.81	3.21	4.49	4.33	5.02
26″	0.020	7.07	3.35	2.95	3.62	5.08	4.52	5.72
28″	0.022	8.04	3.86	3.47	4.14	5.68	4.82	6.39
30″	0.025	8.16	4.20	3.91	4.31	6.23	4.94	7.02
32″	0.026	8.41	4.45	4.35	4.80	6.94	5.13	7.64
34″	0.028	8.98	4.56	4.79	5.31	7.67	5.33	8.26
36″	0.029	9.52	4.74	5.24	5.78	8.35	5.55	8.89
38″	0.031	9.92	5.01	5.47	6.08	8.72	5.73	9.51
40″	0.032	10.31	5.20	6.56	6.38	9.10	5.95	10.13
42″	0.034	10.88	5.29	5.96	6.68	9.48	6.16	10.75
44″	0.036	11.39	5.40	6.20	6.98	9.85	6.39	11.27
46″	0.037	11.73	5.64	6.43	7.28	10.23	6.60	11.79
48″	0.039	12.24	5.84	6.68	7.58	10.61	6.77	12.31
50″	0.041	12.58	5.87	6.92	7.88	10.98	7.01	12.78
52″	0.043	12.75	6.16	7.16.	8.18	11.36	7.18	13.24
54″	0.044	12.92	6.38	7.40	8.48	11.74	7.39	13.71
56″	0.046	13.09	6.52	7.64	8.78	12.11	7.56	14.18
58″	0.048	13.43	6.78	7.88	9.08	12.49	7.86	14.65
60″	0.049	13,60	6.93	8.13	9.38	12.87	8.00	15.11

Low Pressure Spiral Ductwork
Shop Hours
Straight Hours Per Foot; Fittings Hours Per Piece

Automated Fabrication

Diameter	Straight	90 Elbw Offset Wye	45 Elbw Miters Reducer Tap	90T	Lat	90RdT	Sq/Rnd	Cross RdLat LatCross
4"	0.008	0.91	0.47	1.00	1.07	1.70	1.79	1.92
5"	0.008	1.03	0.55	1.02	1.09	1.73	1.86	1.92
6"	0.008	1.18	0.61	1.04	1.10	1.77	1.92	1.92
7"	0.180	1.28	0.67	1.06	1.12	1.80	1.98	1.92
8"	0.009	1.45	0.74	1.08	1.14	1.83	2.05	1.92
9"	0.010	1.58	0.79	1.10	1.24	1.86	2.11	2.05
10"	0.011	1.71	0.87	1.12	1.28	1.90	2.18	2.13
12"	0.012	1.98	1.00	1.16	1.40	2.16	2.33	2.37
14"	0.014	2.24	1.13	1.36	1.53	2.36	2.48	2.60
16"	0.015	2.51	1.27	1.48	1.66	2.57	2.64	2.84
18"	0.017	2.63	1.52	1.61	1.79	2.77	2.80	3.08
20"	0.019	3.35	1.68	1.73	1.91	2.98	2.94	3.31
22"	0.021	4.09	1.90	1.83	2.09	3.17	3.14	3.58
24"	0.023	4.35	2.16	1.99	2.27	3.38	3.26	3.78
26"	0.024	4.99	2.36	2.08	2.56	3.83	3.40	4.31
28"	0.026	5.68	2.72	2.45	2.92	4.28	3.63	4.81
30"	0.029	5.76	2.96	2.76	3.04	4.69	3.72	5.29
32"	0.031	5.93	3.14	3.07	3.39	5.22	3.87	5.75
34"	0.033	6.34	3.22	75.00	3.75	5.77	4.01	6.22
36"	0.034	6.72	3.35	3.70	4.08	6.28	4.18	6.69
38"	0.036	7.00	3.53	3.86	4.29	6.57	4.31	7.16
40"	0.038	7.28	3.67	4.63	4.50	6.85	4.48	7.63
42"	0.040	7.68	3.73	4.21	4.72	7.14	4.64	8.10
44"	0.042	8.04	3.81	4.37	4.93	7.42	4.81	8.49
46"	0.044	8.28	3.98	4.54	5.14	7.70	4.97	8.88
48"	0.046	8.64	4.12	4.72	5.35	7.99	5.10	9.27
50"	0.048	8.88	4.14	4.88	5.56	8.27	5.28	9.62
52"	0.050	9.00	4.35	5.05	5.77	8.55	5.41	9.97
54"	0.052	9.12	4.51	5.23	5.99	8.84	5.56	10.32
56"	0.054	9.24	4.60	5.39	6.20	9.12	5.69	10.68
58"	0.056	9.48	4.79	5.56	6.41	9.40	5.92	11.03
60"	0.058	9.60	4.89	5.74	62	9.69	6.02	11.38

Low Pressure Spiral Ductwork
Field Hours
Straight Hours Per Foot; Fittings Hours Per Piece

Diameter	Straight	90 Elbw 90T Cross	Wye 90RdT	45 Elbw Miters Lat	Reducer Offset RdLat	LatCross Sq/Rd	Tap	Flex Tube
4"	0.054	0.32	0.32	0.32	0.32	0.32	0.20	0.060
5"	0.063	0.36	0.36	0.36	0.36	0.36	0.20	0.070
6"	0.072	0.41	0.41	0.41	0.41	0.41	0.20	0.080
7"	0.072	0.45	0.45	0.45	0.45	0.45	0.20	0.090
8"	0.081	0.50	0.50	0.50	0.50	0.50	0.20	0.100
9"	0.090	0.00	0.00	0.00	0.00	0.00	0.20	0.110
10"	0.099	0.59	0.59	0.59	0.59	0.59	0.20	0.120
12"	0.117	0.59	0.59	0.68	0.68	0.68	0.20	0.140
14"	0.126	0.76	0.76	0.58	0.58	0.58	0.27	0.160
16"	0.135	0.93	0.93	0.71	0.71	0.71	0.34	0.180
18"	0.153	1.11	1.11	0.85	0.85	0.85	0.41	0.190
20"	0.162	1.28	1.28	0.97	0.97	0.97	0.49	0.200
22"	0.180	1.45	1.45	1.11	1.11	1.11	0.56	0.230
24"	0.198	1.62	1.62	1.23	1.23	1.23	0.62	0.260
26"	0.216	1.79	1.79	1.37	1.37	1.37	0.69	0.000
28"	0.234	1.96	1.96	1.50	1.50	1.50	0.77	0.000
30"	0.261	2.14	2.14	1.63	1.63	1.63	0.84	0.000
32"	0.261	2.31	2.31	1.76	1.76	1.76	0.91	0.000
34"	0.288	2.48	2.48	1.90	1.90	1.90	0.98	0.000
36"	0.306	2.66	2.66	2.03	2.03	2.03	1.04	0.000
38"	0.324	2.83	2.83	2.15	2.15	2.15	1.12	0.000
40"	0.333	3.00	3.00	2.28	2.28	2.28	1.19	0.000
42"	0.342	3.18	3.18	2.41	2.41	2.41	1.25	0.000
44"	0.351	3.35	3.35	2.55	2.55	2.55	1.32	0.000
46"	0.369	3.52	3.52	2.68	2.68	2.68	1.40	0.000
48"	0.378	3.69	3.69	2.81	2.81	2.81	1.47	0.000
50"	0.396	3.86	3.86	2.93	2.93	2.93	1.54	0.000
52"	0.414	4.03	4.03	3.06	3.06	3.06	1.61	0.000
54"	0.423	4.21	4.21	3.20	3.20	3.20	1.67	0.000
56"	0.432	4.38	4.38	3.34	3.34	3.34	1.75	0.000
58"	0.441	4.55	4.55	3.47	3.47	3.47	1.82	0.000
60"	0.450	4.73	4.73	3.60	3.60	3.60	1.89	0.000

Medium Pressure Spiral Ductwork
Shop Hours
Straight Hours Per Foot; Fittings Hours Per Piece

Manual Fabrication

Diameter	Straight	90 Elbw Offset Wye	45 Elbw Miters Reducer Tap	90T	Lat	90RdT	Sq/Rnd	Cross RdLat LatCross
4"	0.008	1.52	0.79	1.67	1.78	2.65	1.80	3.00
5"	0.008	1.72	0.92	1.70	1.81	2.70	2.90	3.00
6"	0.008	1.96	1.01	1.73	1.84	2.76	3.00	3.00
7"	0.080	2.13	1.12	1.77	1.87	2.80	3.10	3.00
8"	0.009	2.41	1.23	1.80	1.90	2.86	3.20	3.00
9"	0.010	2.63	1.32	1.84	2.06	2.91	3.30	3.20
10"	0.011	2.85	1.45	1.87	2.13	2.97	3.40	3.33
12"	0.012	3.30	1.67	1.93	2.34	3.37	3.64	3.70
14"	0.014	3.74	1.89	2.26	2.55	3.69	3.88	4.06
16"	0.015	4.19	2.11	2.47	2.76	4.01	4.12	4.43
18"	0.017	4.38	2.54	2.68	2.98	4.33	4.38	4.81
20"	0.019	5.59	2.80	2.89	3.19	4.65	4.60	5.17
22"	0.021	6.82	3.16	3.05	3.49	4.96	4.91	5.59
24"	0.023	7.25	3.60	3.31	3.78	5.28	5.09	5.91
26"	0.024	8.32	3.94	3.47	4.26	5.98	5.32	6.73
28"	0.026	9.46	4.54	4.08	4.87	6.68	5.67	7.52
30"	0.029	9.60	4.94	4.60	5.07	7.33	5.81	8.26
32"	0.031	9.89	5.24	5.12	5.65	8.16	6.04	8.99
34"	0.033	10.56	5.36	5.64	6.25	9.02	6.27	9.72
36"	0.034	11.20	5.58	6.16	6.80	9.82	6.53	10.46
38"	0.036	11.67	5.89	6.44	7.15	10.26	6.74	11.19
40"	0.038	12.13	6.12	7.72	7.50	10.70	7.00	11.92
42"	0.040	12.80	6.22	7.01	7.86	11.15	7.25	12.65
44"	0.042	13.40	6.35	7.29	8.21	11.59	7.52	13.26
46"	0.044	13.80	6.64	7.57	8.56	12.03	7.76	13.87
48"	0.046	14.40	6.87	7.86	8.92	12.48	7.97	14.48
50"	0.048	14.80	6.90	8.14	9.27	12.92	8.25	15.03
52"	0.050	15.00	7.25	8.42	9.62	13.36	8.45	15.58
54"	0.052	15.20	7.51	8.71	9.98	13.81	8.69	16.13
56"	0.054	15.40	7.67	8.99	10.33	14.25	8.89	16.68
58"	0.056	15.80	7.98	9.27	10.68	14.69	9.25	17.23
60"	0.058	16.00	8.15	9.56	11.04	15.14	9.41	17.78

Medium Pressure Spiral Ductwork
Shop Hours
Straight Hours Per Foot; Fittings Hours Per Piece

Automated Fabrication

Diameter	Straight	90 Elbw Offset Wye	45 Elbw Miters Reducer Tap	90T	Lat	90RdT	Sq/Rnd	Cross RdLat LatCross
4"	0.008	1.14	0.59	1.25	1.34	1.99	2.10	2.25
5"	0.008	1.29	0.69	1.28	1.36	2.03	2.18	2.25
6"	0.008	1.47	0.76	1.30	1.38	2.07	2.25	2.25
7"	0.180	1.60	0.84	1.33	1.40	2.11	2.33	2.25
8"	0.009	1.81	0.92	1.35	1.43	2.15	2.40	2.25
9"	0.010	1.97	0.99	1.38	1.55	2.18	2.48	2.40
10"	0.011	2.14	1.09	1.40	1.60	2.23	2.55	2.50
12"	0.012	2.48	1.25	1.45	1.76	2.53	2.73	2.78
14"	0.014	2.81	1.42	1.70	1.91	2.77	2.91	3.05
16"	0.015	3.14	1.58	1.85	2.07	3.01	3.09	3.32
18"	0.017	3.29	1.91	2.01	2.24	3.25	3.29	3.61
20"	0.019	4.19	2.10	2.17	2.39	3.49	3.45	3.88
22"	0.021	5.12	2.37	2.29	2.62	3.72	3.68	4.19
24"	0.023	5.44	2.70	2.48	2.84	3.96	3.82	4.43
26"	0.024	6.24	2.96	2.60	3.20	4.49	3.99	5.05
28"	0.026	7.10	3.41	3.06	3.65	5.01	4.25	5.64
30"	0.029	7.20	3.71	3.45	3.80	5.50	4.36	6.20
32"	0.031	7.42	3.93	3.84	4.24	6.12	4.53	6.74
34"	0.033	7.92	4.02	4.23	4.69	6.77	4.70	7.29
36"	0.034	8.40	4.19	4.62	5.10	7.37	4.90	7.85
38"	0.036	8.75	4.42	4.83	5.36	7.70	5.06	8.39
40"	0.038	9.10	4.59	5.79	5.63	8.03	5.25	8.94
42"	0.040	9.60	4.67	5.26	5.90	8.36	5.44	9.49
44"	0.042	10.05	4.76	5.47	6.16	8.69	5.64	9.95
46"	0.044	10.35	4.98	5.68	6.42	9.02	5.82	10.40
48"	0.046	10.80	5.15	5.90	6.69	9.36	5.98	10.86
50"	0.048	11.10	5.18	6.11	6.95	9.69	6.19	11.27
52"	0.050	11.25	5.44	6.32	7.22	10.02	6.34	11.69
54"	0.052	11.40	5.63	6.53	7.49	10.36	6.52	12.10
56"	0.054	11.55	5.75	6.74	7.75	10.69	6.67	12.51
58"	0.056	11.85	5.99	6.95	8.01	11.02	6.94	12.92
60"	0.058	12.00	6.11	7.17	8.28	11.36	7.06	13.34

Medium Pressure Spiral Ductwork
Field Hours
Straight Hours Per Foot; Fittings Hours Per Piece

Diameter	Straight	90 Elbw 90T Cross	Wye 90RdT	45 Elbw Miters Lat	Reducer Offset RdLat	LatCross Sq/Rd	Tap	Flex Tube
4"	0.060	0.35	0.35	0.35	0.35	0.35	0.22	0.060
5"	0.070	0.40	0.40	0.40	0.40	0.40	0.22	0.070
6"	0.080	0.45	0.45	0.45	0.45	0.45	0.22	0.080
7"	0.080	0.50	0.50	0.50	0.50	0.50	0.22	0.090
8"	0.090	0.55	0.55	0.55	0.55	0.55	0.22	0.100
9"	0.100	0.00	0.00	0.00	0.00	0.00	0.22	0.110
10"	0.110	0.65	0.65	0.65	0.65	0.65	0.22	0.120
12"	0.130	0.65	0.65	0.75	0.75	0.75	0.22	0.140
14"	0.140	0.84	0.84	0.64	0.64	0.64	0.30	0.160,
16"	0.150	1.03	1.03	0.79	0.79	0.79	0.38	0.180
18"	0.170	1.23	1.23	0.94	0.94	0.94	0.46	0.190
20"	0.180	1.42	1.42	1.08	1.08	1.08	0.54	0.200
22"	0.200	1.61	1.61	1.23	1.23	1.23	0.62	0.230
24"	0.220	1.80	1.80	1.37	1.37	1.37	0.69	0.260
26"	0.240	1.99	1.99	1.52	1.52	1.52	0.77	0.000
28"	0.260	2.18	2.18	1.67	1.67	1.67	0.85	0.000
30"	0.290	2.38	2.38	1.81	1.81	1.81	0.93	0.000
32"	0.290	2.57	2.57	1.96	1.96	1.96	1.01	0.000
34"	0.320	2.76	2.76	2.11	2.11	2.11	1.09	0.000
36"	0.340	2.95	2.95	2.25	2.25	2.25	1.16	0.000
38"	0.360	3.14	3.14	2.39	2.39	2.39	1.24	0.000
40"	0.370	3.33	3.33	2.53	2.53	2.53	1.32	0.000
42"	0.380	3.53	3.53	2.68	2.68	2.68	1.39	0.000
44"	0.390	3.72	3.72	2.83	2.83	2.83	1.47	0.000
46"	0.410	3.91	3.91	2.98	2.98	2.98	1.55	0.000
48"	0.420	4.10	4.10	3.12	3.12	3.12	1.63	0.000
50"	0.440	4.29	4.29	3.25	3.25	3.25	1.71	0.000
52"	0.460	4.48	4.48	3.40	3.40	3.40	1.79	0.000
54"	0.470	4.68	4.68	3.56	3.56	3.56	1.86	0.000
56"	0.480	4.87	4.87	3.71	3.71	3.71	1.94	0.000
58"	0.490	5.06	5.06	3.86	3.86	3.86	2.02	0.000
60"	0.500	5.26	5.26	4.00	4.00	4.00	2.10	0.000

This page intentionally left blank

Chapter 12

Estimating Fiberglass Ductwork

INTRODUCTION

Fiberglass ductwork has the following advantages over sheet metal:

- It costs less than conventionally fabricated steel ductwork: 5% less than bare galvanized 30% less than insulated sheet metal 3 5% less than lined sheet metal

- It costs less than galvanized ductwork fabricated on a coil line if insulated or lined: 20% less than insulated 35% less than fined

- However, fiberglass costs 15% more than bare galvanized off of a coil line.

- It requires very little machinery and tools to fabricate, minimal floor space and consequently overhead costs are far less than a sheet metal shop.

- It's more flexible than sheet metal, easier to revise and manipulate at the job site in order to fit into diverse, unpredictable field conditions and to meet precise outlet locations.

- It can be fabricated either in the shop or at the job site.

- It's lightweight. Shipping, material handling and hoisting are far easier than with sheet metal.

- It has built in thermal insulation and acoustical absorption.

- It has built in built in a vapor barrier.

- And lastly, it's an automatic energy saver over bare steel because it's self insulating and virtually leak proof.

Takeoff Procedures
1. Use duct take off sheet and fill out heading.

2. List inside dimensions of duct.

3. Write down type of duct. Use abbreviation.

4. Measure off pipe in 4 foot lengths and list each length.

5. Put down equivalent lengths for each fitting.

6. List sheet metal components separately on bottom of duct take off sheet or on separate sheet, as you go along.

Material Extension
1. Total the linear footage on each line and enter total in total linear foot column.

2. Enter the sq ft/lin ft from the chart which includes the allowance for the 8" comer overlap and a 5% waste factor.

3. Multiply the length times the sq/lin ft and enter in the square feet column.

4. Add up the total square feet.

Material Costs

Typical current prices:

Type Board	0-2400	2400-10,000	Square Feet Ordered Over 10,000 Truckload
475	$0.96	$0.87	$0.79
800	$1.06	$0.10	$0.96
1400	$1.32	$1.28	$1.15

Budget Figures

The budget cost of a 1000 square foot or larger batch of typical fiberglass duct work, under standard conditions, including a markup of 30% for overhead and profit, based on union wages, is about $4.41 per square foot of duct.

Non union cost is about 33% less, roughly $3.04 per square foot.

Square Foot Per Linear Foot of Fiberglass Ductwork

Semi-Perim I.D. Width plus Depth	SF/LF Includes allowances, 8" corner overlap 5% waste	SF/LF No Allowance Included
12	2.8	2.00
14	3.1	2.34
16	3.5	2.67
18	3.9	3.00
20	4.2	3.34
22	4.6	3.67
24	4.9	4.00
26	5.2	4.34
28	5.6	4.67
30	5.9	5.00
32	6.5	5.34
34	6.7	5.67
36	7.0	6.00
38	7.3	6.34
40	7.7	6.67
42	8.0	7.00
44	8.4	7.34
46	8.8	7.67
48	9.1	8.00
50	9.5	8.34
52	9.8	8.67
54	10.1	9.00
56	10.5	9.34
58	10.9	9.67
60	11.2	10.00
62	11.6	10.34
64	11.9	10.67
66	12.3	11.00
68	12.6	11.34
70	12.9	11.67
72	13.3	12.00
74	13.7	12.34
76	14.0	12.67
78	14.4	13.00
80	14.7	13.34
82	15.1	13.67
84	15.4	14.00
86	15.8	14.34
88	16.1	14.67
90	16,5	15.00
92	16.8	15.34
94	17.2	15.67
96	17.5	16.00
98	17.9	16.34
100	18.2	16.67

Calculating Labor

Following labor productivity rates are based on:

1. Volume of 1000 SF and up
2. 800 Board
3. Average duct size of 18" x 12", or equivalent 24 gauge galv.
4. Tie rod reinforcing.
5. Rates based on gross square footage including 8" overlap and 5% waste.
6. Using standard grooving machine.

(See **Fabrication Labor** at the top of the following page.)

Installation Labor

Neither the duct size nor the percentage of fittings affects the installation productivity rate appreciably.

Consequently the erection rate remains relatively constant at 30 sf/hr.

Correction Factors

1. Shop labor
   a. Hand grooving machine versus grooving machine 1.50
   b. Auto grooving machine versus manual machine .... .90
   c. Auto closer versus manual .... .70
   d. Auto groover and closer .... .60
   e. Exhaust and return ducts with tie rods in conduits 1.33
   f. Fittings versus straight duct............................ 1.50
   g. "V" groove versus shiplap 1.05

2. Channel or "T" bar reinforcing versus tie rods
   Shop 1.20
   Field 1.10

3. 475 Board versus 800, Material costs........................ .90
   Shop and Field labor .... .95

4. 1400 Board versus 800, Material costs.................... 1.20
   Shop and Field labor 1.05

5. Material quantity discounts 0-2400 SF.................... 1.05
   2400-20,000 SF 1.00
   Factory, truck load .... .94
   Factory, car load .... .90

FABRICATION LABOR

Percentage	Size Ranges									
	0.12"		13.30"		31.54"		55.84"		85.96"	
Fittings	SF/ HR	HR/ SF	SF/ HR	HR/ SF	SF/ HR	HR/ SF	SF/ HR	HR/ SF	SF/ HR	HR/ SF
Str. Duct. Only	59	.0169	62	.0161	56	.0179	53	.0189	38	.0263
10-20%	55	.0183	58	.0174	52	.0191	49	.0202	35	.0284
20-30%	52	.0191	55	.0182	50	.0200	47	.0211	34	.0297
30-40%	50	.0200	53	.0190	48	.0208	45	.0220	32	.0311
Fittings Only	39	.0256	41	.0244	38	.0263	36	.0278	25	.0400

Example of Fiberglass Ductwork Takeoff and Extension

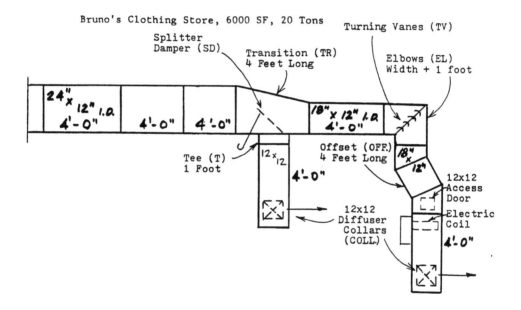

TYPICAL AREA
(Represents 1/10th of job)

DUCT TAKE OFF

Qty Pcs	DUCT SIZE I.D.	TYPE DUCT	Equivalent LINEAR FEET Per Piece	Total LF	SF/LF* Raw	Total SF
3	24 x 12	STR	4 - 4 - 4	12	6	72
1	"	TR	4	4	6	24
2	18 x 12	STR	4 - 4	8	5	40
1	"	EL	2½	2½	5	12½
1	"	OFF	4	4	5	20
1	12 x 12	T	1	1	4	4
1	"	STR	4	4	4	16

10 Pcs   Totals   188.5 SF Raw

Allowances: Add 8" per linear foot for overlap 24 = .66 x 35.5 LF

         195

Add 5% for waste   14

Gross total for 1 area   222.5 SF

     Times 10 areas   x 10

Job total   2225 SF

*Can also use SF/LF from chart which has the 8" corner allowance and 5" waste built in already.

Pricing Sheet Metal Components
Fiberglass Ductboard

ITEM	SIZE	DIRECT MATERIAL & LABOR
Turning Vanes	12x6	$24.24
	18x12	50.10
	24x12	58.18
	36x12	82.42
	42x12	111.52
	48x24	153.54
	60x24	205.25
Splitter Dampers	9x9	$30.71
	18x12	35.55
	30x12	43.63
	42x18	54.95
Round Diffuser Collar	8" dia	$54.95
	12	71.11
	18	82.42
Rectangular Diffuser Collar	12x12	$69.50
	24x12	82.42
	48x24	187.47
Fire Damper Sleeves 18 Gauge	24x12	$143.84
	36x18	190.71
	48x12	260.20
Electric Coil Sleeves	12x12	$74.34
	24x12	108.28
	36x12	143.84
	48x12	205.25
Acces Doors	12x12	$72.73
	24x12	87.28
Reinforcing Channels and formed "T" Bars	—Material Labor-Fab & Install	$0.20LF
	24" long	.10 Hrs ea
	36"	.14
	48"	.18
	60"	.22
	72"	.26

Estimate Summary and Extension Sheet

Job **Bruno's Clothing Store**

FIBER GLASS DUCTWORK TYPE 800

		MATERIAL COST UNIT	MATERIAL COST TOTAL	SHOP HRS UNIT	SHOP HRS TOTAL	FIELD HRS UNIT	FIELD HRS TOTAL
Fiber-Glass Ductboard, 2090 Sq.Ft.		$.84 SF	$1,756	.55 SF/HR	38	.30 SF/HR	70
Accessories, Tape,							
Hangers, etc. 2090 sq.ft.		$.16/SE	334				
Shop Drawings "				400 SF/HR	5		
Cartage "				1000 SF/HR	2		
	Ea.						
Splitter Dampers, 12x12	10	$25	$250	(Labor Included)			
Turning Vanes, 12x12	10	45	450				
Diffuser Collars, 12x12	20	59	1,180				
Electric Coil Sleeves, 18x12	10	75	750				
Access Doors, 12x12	10	60	600				
			$5,320		45 HRS		70 HRS
Total Direct Costs			+$3,795		70		
30% O+P			9,115	← 115 HRS @ $33.00			
Total Sell			+ 2,735				
			$ 11,850				

Cost Per Square Foot Breakdown   Union $33/hr   Non Union @ $24/hr

	Union	Non Union
E.G. Board #800	$.84/SF	.84/SF
Accessories	.20	.20
Shop Labor, 55 SF/HR	.60	.44
Field Labor, 30 SF/HR	1.10	.80
Shop Drawings, 400 SF/HR	.08	.06
Cartage, 1000 SF/HR	.03	.03
Direct Costs	2.85	2.37
30% O.+P.	.86	.71
	$3.71 sq.ft	3.08 sq.ft

Chapter 13

Heavy Gauge Ductwork

TYPES OF INDUSTRIAL EXHAUST DUCTWORK

There are four basic categories of materials used for ductwork in air pollution control and, industrial and commercial exhaust systems.

1. **Galvanized** ductwork, either round or rectangular, is used in many applications for heat, moisture and dust removal when corrosion and abrasion do not present problems.

2. **Black iron** ductwork, hot or cold rolled, round or rectangular is used in many applications for heat, moisture and dust removal when corrosion and abrasion do not present problems.

3. **Corrosion and moisture resistant** ductwork round or rectangular, is fabricated from the following materials:

PVC	Stainless Steel
FRP	PVC Coated Galvanized
Aluminum	Transite

4. **High temperature** ductwork is generally stainless steel.

Round ductwork is used more commonly than rectangular in air pollution control and material conveying work. Connections may be cleated, slip, sheet metal flanged or companion angle flanged and bolted, welded, soldered, cemented, gasketed, sleeved, coupled, or riveted.

Budget Estimating
Rectangular Black Iron Ductwork
Metal Flanges
Per Pound and Per Square Foot

Average Lb/Hr, Sf/Hr Labor Productivity Rates and Budget Prices

Gauge	Fabrication		Installation		Budget Price	W/O & P
	Lb/Hr	Sf/Hr	Lb/Hr	Sf/Hr	Per/Lb	Per/Sf
18	28	14	34	17.0	$4.73	$ 9.44
16	31	12.5	39	15.5	4.32	10.60
14	35	11	45	14.2	3.57	11.52
12	44	10	50	11.5	3.43	15.13
10	52	9	55	10.0	3.05	16.73
3/16"	53	7.0	41	5.4	3.30	25.45

Based on $0.40 per pound for hot rolled steel and $39.00 for labor. Labor productivity based on average 6 foot perimeter (24x12 plus or minus) and 25 percent fittings by weight or square footage.

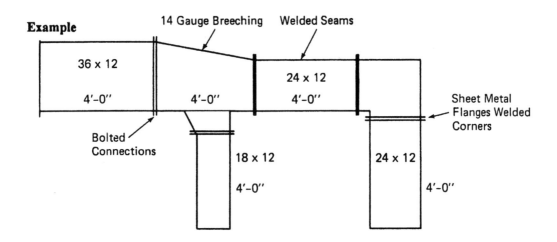

Example — 14 Gauge Breeching — Welded Seams

36 x 12 — 4'-0"
24 x 12 — 4'-0"
Bolted Connections
18 x 12 — 4'-0"
24 x 12 — 4'-0"
Sheet Metal Flanges Welded Corners

	Size	SF/LF	Linear Feet	Total LF	Square Feet	
2	36x12	8	4-4	8	64	
3	24x12	6	4-3-4	11	66	
2	18x12	5	1-4	5	25	
7 pcs			Totals	24 LF	155 SF	
			@ 3.125 LBS/SF		485 LBS	
			Waste 30%		145	
					630 LBS	
			Material @	31¢/LB	$195.30	
			Shop @ 35 LBS/HR		18 HRS	
			Field @ 45 LBS/HR		14 HRS	

Heavy Gauge Ductwork Labor Factors
Showing Relationship Between Different Materials, Gauges and Connections

TYPE OF MATERIAL LABOR FACTOR

	Multipliers	
	SHOP	**FIELD**
Black Iron, (Base)	1.00	1.00
Stainless Steel	1.25	1.25
Galvanized	1.10	1.06
Aluminum	.75	.75
PVC 1/4", 3/16"	.43	.82
FRP 1/4", 3/16"	.83	.87
Aluminized Steel	1.00	1.00
Cor-Ten	1.00	1.00

ANGLE FLANGE CONNECTION, BLACK IRON
Gauge Factors

| GAUGE | Multipliers | |
	SHOP	FIELD
20	.75	.75
18	.80	.80
16	.90	.90
14 (Base)	1.00	1.00
12	1.10	1.25
10	1.20	1.45
8	1.40	2.00
3/16″	1.60	2.60
1/4″)	2.13	3.40
3/8″	2.63	3.90
1/2″	3.13	4.40

BENT METAL FLANGE CONNECTION
(As Compared to Angle Flange Connection)

Shop Factor	.90	Multipliers
Field Factor	.95	

WELDED CONNECTION
(As Compared to Angle Flange Connection)

Shop Factor	.60	Multipliers
Field Factor	1.50	

Examples: Given 10 gauge black iron duct with angle flanges shop… 200/0 more labor, field 45% more. Given same gauge duct with welded connections shop 40% less labor, field 50% more.

Rectangular Black Iron Ductwork
With Companion Angles
Per Piece Labor and Material

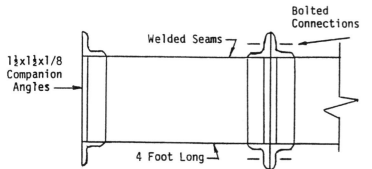

Uses: Boiler breechings, oven exhausts, material conveying, kitchens

Correction Factors (On 14 gauge per piece labor)

		Shop	Field
1.	18 gauge	.80	.80
2.	16 gauge	.90	.90
3.	12 gauge	1.10	1.25
4.	10 gauge	1.20	1.45
5.	3/16ths inch plate	1.60	2.60
6.	5 foot long joints	1.10	1.10
7.	8 foot long joints	1.50	1.50

Budget price, typical 14 gauge 24x12 duct section. $3.49 per lb.

Rectangular Heavy Gauge Ductwork
14 Gauge Black Iron, Angle Flange Connection
Shop Hours
Fittings Hours Per Piece; Straight Hours Per Foot

Manual Fabrication

Size	Straight	90 Elbw	45 Elbw	Sq/Elbw Trans	Tap	Wye	Sq/Rd	Offset	Door
12x12	2.24	3.45	2.58	2.58	2.04	3.04	6.49	3.87	2.73
18x12	2.58	4.13	3.10	3.00	2.40	3.59	7.20	4.50	2.86
24x12	2.92	4.81	3.61	3.42	2.76	4.14	7.90	5.13	2.99
30x12	3.25	5.49	4.12	3.84	3.12	4.69	8.61	5.76	3.11
36x12	3.59	6.17	4.63	4.26	3.48	5.24	9.38	6.39	3.24
36x18	3.92	6.85	5.14	4.68	3.84	5.80	10.28	7.02	3.36
42x18	4.26	8.64	6.48	5.10	4.20	6.35	11.18	7.65	3.49
48x18	4.60	9.43	7.08	5.52	4.56	6.90	12.08	8.28	3.61
54x18	4.93	10.22	7.67	5.94	4.92	7.45	12.97	8.91	3.74
60x15	5.27	11.02	8.27	6.36	5.28	8.00	13.87	9.54	3.86
72x18	5.94	16.49	12.37	7.20	6.00	9.11	15.67	10.80	4.11
84x18	6.61	18.59	13.95	8.04	6.72	10.21	17.46	12.06	4.36

Automated Fabrication

Size	Straight	90 Elbw	45 Elbw	Sq/Elbw Trans	Tap	Wye	Sq/Rd	Offset	Door
12x12	1.68	2.59	1.94	1.94	1.53	2.28	4.87	2.90	2.05
18x12	1.94	3.10	2.33	2.25	1.80	2.69	5.40	3.38	2.15
24x12	2.19	3.61	2.71	2.57	2.07	3.11	5.93	3.85	2.24
30x12	2.44	4.12	3.09	2.88	2.34	3.52	6.46	4.32	2.33
36x12	2.69	4.63	3.47	3.20	2.61	3.93	7.04	4.79	2.43
36x18	2.94	5.14	3.86	3.51	2.88	4.35	7.71	5.27	2.52
42x18	3.20	6.48	4.86	3.83	3.15	4.76	8.39	5.74	2.62
48x18	3.45	7.07	5.31	4.14	3.42	5.18	9.06	6.21	2.71
54x18	3.70	7.67	5.75	4.46	3.69	5.59	9.73	6.68	2.81
60x18	3.95	8.27	6.20	4.77	3.96	6.00	10.40	7.16	2.90
72x18	4.46	12.37	9.28	5.40	4.50	6.83	11.75	8.10	3.08
84x18	4.96	13.94	10.46	6.03	5.04	7.66	13.10	9.05	3.27

Rectangular Heavy Gauge Ductwork
14 Gauge Black Iron, Angle Flange Connection
Field Hours
Fittings Hours Per Piece; Straight Hours Per Foot

Size	Straight	45 Elbw 90 Elbw Wye	Sq/Elbw Trans Offset	Tap	Sq/Rd
12x12	1.60	1.44	1.60	0.39	1.16
18x12	1.90	1.71	1.90	0.46	1.35
24x12	2.20	1.98	2.20	0.53	1.55
30x12	2.50	2.25	2.50	0.60	1.74
36x12	2.80	2.52	2.80	0.68	1.93
36x18	3.10	2.79	3.10	0.75	2.12
42x18	3.40	3.06	3.40	0.82	2.31
46x18	3.70	3.33	3.70	0.89	2.50
54x18	4.00	3.60	4.00	0.96	2.69
60x18	4.30	3.87	4.30	1.04	2.88
72x18	4.90	4.41	4.90	1.18	3.26
84x18	5.50	4.95	5.50	1.32	3.65

Rectangular Heavy Gauge Ductwork
14 Gauge Galvanized, Angle Flange Connection
Shop Hours 4
Fittings Hours Per Piece; Straight Hours 4 Foot

Manual Fabrication

Size	Straight	90 Elbw	45 Elbw	Sq/Elbw Trans	Tap	Wye	Sq/Rd	Offset	Door
12x12	2.46	3.80	2.84	2.84	2.24	3.34	7.14	4.26	3.00
18x12	2.84	4.54	3.41	3.30	2.64	3.95	7.92	4.95	3.15
24x12	3.21	5.29	3.97	3.76	3.04	4.55	8.69	5.64	3.29
30x12	3.58	6.04	4.53	4.22	3.43	5.16	9.47	6.34	3.42
36x12	3.95	6.79	5.09	4.69	3.83	5.76	10.32	7.03	3.56
36x18	4.31	7.54	5.65	5.15	4.22	6.38	11.31	7.72	3.70
42x18	4.69	9.50	7.13	5.61	4.62	6.99	12.30	8.42	3.84
48x18	5.06	10.37	7.79	6.07	5.02	7.59	13.29	9.11	3.97
54x18	5.42	11.24	8.44	6.53	5.41	8.20	14.27	9.80	4.11
60x18	5.80	12.12	9.10	7.00	5.81	8.80	15.26	10.49	4.25
72x18	6.53	18.14	13.61	7.92	6.60	10.02	17.24	11.88	4.52
84x18	7.27	20.45	15.35	8.84	7.39	11.23	19.21	13.27	4.80

Automated Fabrication

Size	Straight	90 Elbw	45 Elbw	Sq/Elbw Trans	Tap	Wye	Sq/Rd	Offset	Door
12x12	1.85	2.85	2.13	2.13	1.68	2.51	5.35	3.19	2.25
18x12	2.13	3.41	2.56	2.48	1.98	2.96	5.94	3.71	2.36
24x12	2.41	3.97	2.98	2.82	2.28	3.42	6.52	4.23	2.47
30x12	2.68	4.53	3.40	3.17	2.57	3.87	7.10	4.75	2.57
36x12	2.96	5.09	3.82	3.51	2.87	4.32	7.74	5.27	2.67
36x18	3.23	5.65	4.24	3.86	3.17	4.79	8.48	5.79	2.77
42x18	3.51	7.13	5.35	4.21	3.47	5.24	9.22	6.31	2.88
48x18	3.80	7.78	5.84	4.55	3.76	5.69	9.97	6.83	2.98
54x18	4.07	8.43	6.33	4.90	4.06	6.15	10.70	7.35	3.09
60x18	4.35	9.09	6.82	5.25	4.36	6.60	11.44	7.87	3.18
72x18	4.90	13.60	10.21	5.94	4.95	7.52	12.93	8.91	3.39
84x18	5.45	15.34	11.51	6.63	5.54	8.42	14.40	9.95	3.60

Rectangular Heavy Gauge Ductwork
14 Gauge Galvanized, Angle Flange Connection
Field Hours
Fittings Hours Per Piece; Straight Hours Per Foot

Size	Straight	45 Elbw 90 Elbw Wye	Sq/Elbw Trans Offset	Tap	Sq/Rd
12x12	1.60	1.44	1.60	0.39	1.16
18x12	1.90	1.71	1.90	0.46	1.35
24x12	2.20	1.98	2.20	0.53	1.55
30x12	2.50	2.25	2.50	0.60	1.74
36x12	2.80	2.52	2.80	0.68	1.93
36x18	3.10	2.79	3.10	0.75	2.12
42x18	3.40	3.06	3.40	0.82	2.31
48x18	3.70	3.33	3.70	0.89	2.50
54x18	4.00	3.60	4.00	0.96	2.69
60x18	4.30	3.87	4.30	1.04	2.88
72x18	4.90	4.41	4.90	1.18	3.26
84x18	5.50	4.95	5.50	1.32	3.65

Round Black Iron Ductwork
Fabrication Labor Hours Per Piece
14 and 16 Gauge Welded

Manual Fabrication

DIA.	STR. PIPE 4 FT Long	90° ELBOWS Gores			45°,60° ELBOWS 3 Gores	45° TEE	45° REDUC- ING TEE	90° TEE	90° REDUC- ING TEE	SQ. to RND.	WELD ANGLE RINGS ON EA.
		5	4	3							
4	.7	2.0	1.7	1.4	1.0	2.3	3.7	2.1	3.4	3.5	.50
5	.7	2.3	2.0	1.6	1.2	2.3	3.7	2.1	3.4	3.6	.51
6	.7	2.5	2.1	1.7	1.3	2.3	3.7	2.1	3.4	3.7	.52
7	.7	2.6	2.2	1.8	2.3	2.4	3.8	2.2	3.5	3.9	.54
8	.7	3.0	2.6	2.1	1.5	2.4	3.8	2.2	3.5	4.0	.55
9	.8	3.3	2.8	2.3	1.7	2.6	4.1	2.3	3.7	4.2	.57
10	.8	3.5	3.0	2.5	1.8	2.6	4.1	2.3	3.7	4.3	.58
11	.8	3.8	3.2	2.7	1.9	2.7	4.3	2.4	3.8	4.6	.59
12	.9	4.0	3.4	2.9	2.0	2.9	4.6	2.6	4.2	4.8	.60
14	.9	4.7	4.0	3.3	2.4	3.2	5.1	2.9	4.6	5.0	.64
16	.9	5.4	4.6	3.8	2.7	3.4	5.4	3.1	5.0	5.2	.69
18	1.1	6.0	5.1	4.2	3.0	3.8	6.1	3.4	5.4	5.3	.74
20	1.1	7.0	6.0	4.9	3.5	4.0	6.4	3.6	5.8	5.7	.78
22	1.2	8.0	6.7	5.6	4.0	4.3	6.9	3.9	6.2	5.9	.86
24	1.3	9.0	7.7	6.3	4.5	4.7	7.5	4.1	6.6	6.3	.95
26	1.4	10.0	8.5	7.0	5.0	5.3	8.5	4.8	7.7	6.6	1.01
28	1.5	11.0	9.4	7.7	5.5	5.8	9.3	5.2	8.3	6.8	1.06
30	1.6	12.0	10.2	8.4	6.0	6.4	10.2	5.8	9.3	7.0	1.12
36	2.7	14.0	10.9	9.8	7.0	8.5	13.6	7.7	12.3	8.0	1.34
42	3.0	16.0	13.6	11.2	8.0	9.7	15.5	8.7	13.9	9.0	1.52
48	3.3	18.0	15.3	12.6	9.0	11.3	18.1	9.8	15.7	10.0	1.75

Standard pipe lengths 4 ft. from standard mill metal widths.

Correction Factors on Per Piece Labor
1. 7 gore 90* elbow..........................1.30
2. 10 gauge..1.25
3. 12 gauge..1.15
4. 18 gauge...90
5. 20 gauge...85
6. Galvanized.....................................1.15
7. Welded together on floor.............60
8. Duplicates after first.....................66

BLOW PIPE GAUGES

Diameter of Straight Ducts	Non Abrasive Class I	Medium Abrasive Class II	Highly Abrasive Class III
to 8"	24	22	20
over 8" to 18"	22	20	18
over 18" to 30"	20	18	16
over 30"	18	16	14

Round Black Iron Ductwork
16 Gauge, Shop Hours
Fittings Hours Per Piece; Straight Hours Per Foot

Automated Fabrication

Diameter	Straight	Sq/Elbw 45RE	Trans	90RE	Tap	Wye	Sq/Rnd	Offset	2 Rings
4	0.138	0.806	0.597	1.612	0.377	2.015	2.319	0.896	1.480
5	0.145	0.882	0.633	1.763	0.395	2.204	2.417	0.949	1.500
6	0.153	0.957	0.668	1.914	0.413	2.393	2.514	1.003	1.540
7	0.161	1.033	0.704	2.066	0.431	2.582	2.611	1.056	1.600
8	0.168	1.108	0.739	2.217	0.450	2.771	2.709	1.109	1.640
9	0.176	1.184	0.775	2.368	0.468	2.960	2.806	1.162	1.700
10	0.184	1.260	0.810	2.519	0.486	3.149	2.904	1.215	1.740
12	0.199	1.411	0.881	2.822	0.522	3.527	3.098	1.322	1.800
14	0.214	1.632	0.952	3.265	0.559	4.081	3.293	1.428	1.940
16	0.229	1.886	1.023	3.772	0.595	4.716	3.488	1.534	2.100
18	0.245	2.140	1.093	4.280	0.632	5.350	3.683	1.641	2.240
20	0.260	2.394	1.164	4.788	0.668	5.985	3.878	1.747	2.380
22	0.275	2.648	1.235	5.295	0.705	6.619	4.072	1.854	2.540
24	0.291	2.901	1.306	5.803	0.741	7.254	4.267	1.960	2.840
26	0.306	3.155	1.377	6.310	0.778	7.888	4.462	2.066	3.040
28	0.321	3.409	1.448	6.818	0.814	8.523	4.657	2.173	3.240
30	0.337	3.663	1.519	7.326	0.850	9.157	4.852	2.279	3.440
32	0.549	3.917	1.590	7.833	0.887	9.792	4.923	2.386	3.680
34	0.575	4.170	1.661	8.341	0.923	10.426	5.161	2.492	3.940
36	0.602	4.424	1.731	8.848	0.960	11.061	5.399	2.598	4.140
38	0.628	4.678	1.802	9.356	0.996	11.695	5.637	2.705	4.380
40	0.655	4.932	1.873	9.864	1.033	12.330	5.875	2.811	4.620
42	0.681	5.186	1.944	10.371	1.069	12.964	6.113	2.917	4.840
44	0.708	5.439	2.015	10.879	1.106	13.599	6.351	3.024	5.140
46	0.734	5.693	2.086	11.386	1.142	14.233	6.589	3.130	5.440
48	0.761	5.947	2.157	11.894	1.179	14.868	6.827	3.237	5.700
60	0.787	6.201	2.227	12.402	1.215	15.502	7.065	3.343	5.940
64	0.841	6.708	2.369	13.417	1.288	16.771	7.541	3.556	6.480
58	0.894	7.216	2.511	14.432	1.361	18.040	8.017	3.768	7.000
62	0.947	7.724	2.653	15.447	1.434	19.309	8.493	3.981	7.560
66	1.000	8.231	2.794	16.462	1.507	20.578	8.969	4.194	8.240
70	1.053	8.739	2.936	17.478	1.580	21.847	9.445	4.407	8.640
74	1.106	9.246	3.078	18.493	1.652	23.116	9.921	4.619	9.100
78	1.159	9.754	3.220	19.508	1.725	24.385	10.397	4.832	9.500
82	1.212	10.262	3.361	20.523	1.798	25.654	10.874	5.045	9.960
86	1.265	10.769	3.503	21.538	1.871	26.923	11.350	5.258	10,400
90	1.319	11.277	3.645	22.554	1.944	28.192	11.826	5.471	11.000
94	1.372	11.784	3,787	23.569	2.017	29.461	12.302	5.683	11.600
98	1.425	12.292	3.928	24.584	2.090	30.730	12.778	5.896	12.200

Note: Angle ring labor includes welding time to attach rings to each end of piece.

Round Black Iron Ductwork
16 Gauge, Shop Hours
Fittings Hours Per Piece; Straight Hours Per Foot

Automated Fabrication

Diameter	90T	90RdT	90Crs	Lateral	RdLat	LatCrs	Reducer	End Cap	Ac Door
4	1.100	1.880	2.225	1.156	1.976	2.194	1.109	0.675	1.156
5	1.156	1.975	2.336	1.226	2.094	2.324	1.192	0.675	1.578
6	1.213	2.071	2.448	1.298	2.215	2.456	1.275	0.675	1.592
7	1.272	2.169	2.562	1.371	2.338	2.591	1.358	0.675	1.606
8	1.331	2.268	2.678	1.445	2.462	2.727	1.441	0.675	1.620
9	1.391	2.369	2.796	1.520	2.589	2.866	1.524	0.675	1.634
10	1.452	2.472	2.915	1.597	2.718	3.007	1.606	0.675	1.648
12	1.578	2.682	3.159	1.754	2.981	3.295	1.772	0.675	1.677
14	1.707	2.898	2.482	1.916	3.253	1.709	1.710	0.675	1.705
16	1.840	3.120	2.976	2.083	3.533	2.227	1.960	0.675	1.733
18	1.714	2.904	3.487	2.256	3.822	2.762	2.211	0.675	1.761
20	1.977	3.346	4.014	2.433	4.10	3.314	2.462	0.898	1.789
22	2.249	3.802	4.557	2.386	4.043	3,884	2.712	0.898	1.818
24	2.530	4.272	5.115	2.710	4.578	4.471	2.963	0.898	1.846
26	2.819	4.757	5.690	3.045	5.138	5.076	3.214	0.898	1.874
28	3.117	5.255	6.281	3.390	5.714	5.697	3.464	0.898	1.902
30	3.424	5.767	6.888	3.745	6.308	6.337	3.715	0.898	1.931
32	3.740	6.294	7.511	4.111	6.918	6.993	3.966	0.898	1.959
34	4.065	6.835	8.150	4.486	7.544	7.667	4.216	0.898	1.987
36	4.398	7.389	8.805	4.873	8.187	8.358	4.467	0.898	2.015
38	4.741	7.958	9.476	5.269	8.846	9.067	4.718	1.120	2.043
40	5.092	8.541	10.163	5.676	9.522	9.793	4.969	1.120	2.072
42	5.452	9.138	10.867	6.094	10.215	10.537	5.219	1.120	2.100
44	5.821	9.749	11.586	6.521	10,924	11.297	5.470	1.120	2.128
46	6.198	10.375	12.321	6.959	11.649	12.076	5.721	1.120	2.156
48	6.585	11.014	13.073	7.408	12.392	12.871	5.971	1.120	2.184
50	6.980	11.668	13.840	7.866	13.150	13.684	6.222	1.120	2.213
54	7.797	13.017	15.423	8.815	14.717	15.362	6.723	1.120	2.269
58	8.649	14.423	17.070	9.804	16.351	17.109	7.225	1.350	2.362
62	9.536	15.885	18.782	10.835	18.050	18.926	7.726	1.350	2.382
66	10.459	17.404	20.557	11.908	19.816	20.813	8.228	1.350	2.438
70	11.417	18.979	22.397	13.022	21.648	22.769	8.729	1.350	2.495
74	12.411	20.611	24.301	14.177	23.546	24.794	9.230	1.573	2.551
78	13.439	22.299	26.270	15.373	25.510	26.889	9.732	1.573	2.608
82	14.503	24.044	28.302	16.611	27.540	29.054	10.233	1.573	2.664
86	15.602	25.845	30.399	17.890	29.637	31.288	10.735	1.573	2.721
90	16.737	27.702	32.560	19.211	31.799	33.591	11.236	1.573	2.777
94	17.906	29.616	34.785	20.573	34.028	35.964	11.737	1.796	2.833
98	19.111	31.586	37.074	21.976	36.323	38.407	12.239	2.700	2.890

Note: Transition and reducer hours based on 4' lengths.

Round Black Iron Ductwork
16 Gauge, Angle Ring Connection, Installation Labor
Fittings Hours Per Piece; Straight Hours Per Foot

Diameter	Straight	45RE 90RE Wye	Sq/Rnd Sq/Elbw Trans	Tap	Offset
4	0.151	0.544	0.540	0.133	0.540
5	0.179	0.643	0.585	0.144	0.585
6	0.206	0.743	0.630	0.155	0.630
7	0.234	0.842	0.675	0.166	0.675
8	0.262	0.942	0.720	0.176	0.720
9	0.289	1.042	0.765	0.187	0.765
10	0.317	1.141	0.810	0.198	0.810
12	0.372	1.341	0.900	0.220	0.900
14	0.428	1.640	0.990	0.241	0.990
16	0.483	1.739	1.080	0.263	1.080
18	0.538	1.938	1.170	0.284	1.170
20	0.594	2.138	1.260	0.306	1.260
22	0.649	2.337	1.350	0.328	1.350
24	0.704	2.536	1.440	0.349	1.440
26	0.760	2.735	1.530	0.371	1.530
28	0.815	2.935	1.620	0.392	1.620
30	0.871	3.134	1.710	0.414	1.710
32	0.926	3.333	1.800	0.436	1.800
34	0.981	3.532	1.890	0.457	1.890
36	1.037	3.732	1.980	0.479	1.980
38	1.092	3.931	2.070	0.500	2.070
40	1.147	4.130	2.160	0.522	2,160
42	1.203	46329	2.250	0.544	2.250
44	1.258	4.529	2.340	0.565	2.340
46	1.313	4.728	2.430	0.587	2.430
48	1.369	4.927	2.520	0.608	2.520
50	1.424	5.126	2.610	0.630	2.160
54	1.535	5.525	2.790	0.673	2.790
58	1.645	5.924	2.970	0.716	2.970
62	1.756	6.322	3.150	0.760	3.150
66	1.867	6.721	3.330	0.803	3.330
70	1.978	7.119	3.510	0.846	3.510
74	2.088	7.518	3.690	0.889	3.690
78	2.199	7.916	3.870	0.932	3.870
82	2.310	8.315	4.050	0.976	4.050
86	2.420	8.713	4.230	1.019	4.230
90	2.531	9.112	4.410	1.062	4.410
94	2.642	9.510	4.590	1.105	4.590
98	2.752	9.909	4.770	1.148	4.770

Note: Angle ring connection labor is included with per piece labor.

Round Black Iron Ductwork
16 Gauge, Angle Ring Connection, Installation Labor
Fittings Hours Per Piece; Straight Hours Per Foot

Diameter	90T Lateral	90RdT RdLat	90Crs LatCrs	Reducer	BlastGt
4	0.443	0.473	0.503	0.604	0.900
5	0.530	0.566	0.601	0.715	0.956
6	0.619	0.660	0.702	0.825	1.012
7	0.710	0.757	0.803	0.936	1.069
8	0.802	0.855	0.907	1.047	1.125
9	0.897	0.955	1.013	1.157	1.181
10	0.993	1.057	1.120	1.268	1.237
12	1.192	1.266	1.341	1.490	1.350
14	1.397	1.483	1.568	1.711	1.462
16	1.610	1.707	1.803	1.932	1.575
18	1.831	1.938	2.046	2.154	1.688
20	2.058	2.171	2.296	2.375	1.800
22	2.294	2.423	2.553	2.596	1.912
24	2.536	2.677	2.818	2.818	2.025
26	2.786	2.938	3.090	3.039	2.138
28	3.043	3.206	3.369	3.261	2.250
30	3.308	3.482	3.656	3.482	2.362
32	3.580	3.765	3.950	3.704	2.475
34	3.859	4.056	4.252	3.925	2.587
36	4.146	4.354	4.561	.4.146	2.700
38	4.440	4.659	4.877	4.368	2.813
40	4.742	4.972	5.201	4.589	2.925
42	5.051	5.292	5.532	4.811	3.037
44	5.367	5.619	5.871	5.032	3.150
46	5.691	5.954	6.216	5.235	3.262
48	6.022	6.296	6.570	5.475	3.375
50	6.361	6.645	6.930	5.696	3.487
54	7.060	7.367	7.674	6.139	3.712
58	7.788	8.117	8.447	6.582	3.938
62	8.546	8.898	9.249	7.024	4.162
66	9.334	9.707	10.081	7.467	4.387
70	10.151	10.547	10.942	7.910	4.612
74	10.998	11.416	11.833	8.353	4.838
78	11.874	12.314	12.754	8.796	5.063
82	12.780	13.242	13.704	9.238	5.287
86	13.715	14.199	14.683	9.681	5.512
90	14.680	15.186	15.692	10.124	5.737
94	15.674	16.203	16.731	10.567	5.962
98	16.698	17.249	17.799	11.010	6.188

Note: Angle ring connection labor is included with per piece labor. Reducer labor based on 4' lengths.

Rolled Steel Angle Rings

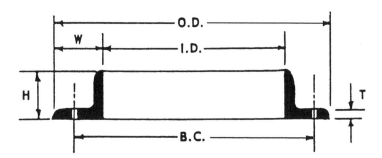

TABLE OF STANDARD SIZES

Nominal Size	I.D. Actual Inside Diameter Inches	O.D. Outside Diameter Inches	Angle Size Inches H x W x T	B.C. Bolt Circle Inches	Holes Size Inches	Holes Number	Approx Weight Lbs Each	Approx Cost Each
4	4-1/8	6-1/8	1 x 1 x 1/8	5-1/4	9/32	6	1.00	
5	5-1/8	7-1/8	1 x 1 x 1/8	6-5/16	9/32	6	1.25	$5.88
6	6-1/8	8-5/8	1-1/4 x 1-1/4 x 1/8	7-1/2	11/32	6	1.75	
7	7-1/8	9-5/8	1-1/4 x 1-1/4 x 1/8	8-1/2	11/32	6	2.00	
8	8-1/8	10-5/8	1-1/4 x 1-1/4 x 1/8	9-1/2	11/32	8	2.25	$12.97
9	9-1/8	11-5/8	1-1/4 x 1-1/4 x 1/8	10-1/2	11/32	8	2.50	
10	10-3/16	12-11/16	1-1/4 x 1-1/4 x 1/8	11-9/16	11/32	8	2.75	
11	11-3/16	13-11/16	1-1/4 x 1-1/4 x 1/8	12-9/16	11/32	8	3.00	
12	12-3/16	15-3/16	1-1/2 x 1-1/2 x 3/16	13-13/16	13/32	12	6.00	
13	13-3/16	16-3/16	1-1/2 x 1-1/2 x 3/16	14-13/16	13/32	12	6.50	
14	14-3/16	17-3/16	1-1/2 x 1-1/2 x 3/16	15-13/16	13/32	12	7.00	$14.16
15	15-3/16	18-3/16	1-1/2 x 1-1/2 x 3/16	16-13/16	13/32	16	7.50	
16	16-1/4	19-3/4	1-3/4 x 1-3/4 x 3/16	18-1/8	13/32	16	9.50	
17	17-1/4	20-3/4	1-3/4 x 1-3/4 x 3/16	19-1/8	13/32	16	10.00	
18	18-1/4	21-3/4	1-3/4 x 1-3/4 x 3/16	20-1/8	13/32	16	10.50	$18.86
19	19-1/4	22-3/4	1-3/4 x 1-3/4 x 3/16	21-1/8	13/32	16	11.25	
20	20-1/4	23-3/4	1-3/4 x 1-3/4 x 3/16	22-1/8	13/32	20	11.75	
21	21-1/4	24-3/4	1-3/4 x 1-3/4 x 3/16	23-1/8	9/16	20	12.25	
22	22-1/4	25-3/4	1-3/4 x 1-3/4 x 3/16	24-1/8	9/16	20	12.75	
23	23-1/4	26-3/4	1-3/4 x 1-3/4 x 3/16	25-1/8	9/16	20	13.50	$23.61
24	24-1/4	27-3/4	1-3/4 x 1-3/4 x 3/16	26-1/8	9/16	20	14.00	
25	25-1/4	29-1/4	2 x 2 x 3/16	27-1/2	9/16	20	16.50	
26	26-1/4	30-1/4	2 x 2 x 3/16	28-1/2	9/16	24	17.25	
27	27-1/4	31-1/4	2 x 2 x 3/16	29-1/2	9/16	24	18.00	$29.26
28	28-1/4	32-1/4	2 x 2 x 3/16	30-1/2	9/16	24	18.50	
29	29-1/4	33-1/4	2 x 2 x 3/16	31-1/2	9/16	24	19.25	
30	30-1/4	34-1/4	2 x 2 x 3/16	32-1/2	9/16	28	20.00	
32	32-1/4	36-1/4	2 x 2 x 3/16	34-1/2	9/16	28	21.25	
34	34-1/4	38-1/4	2 x 2 x 3/16	36-1/2	9/16	32	22.50	$35.39
36	36-1/4	40-1/4	2 x 2 x 3/16	38-1/2	9/16	32	23.75	
38	38-1/4	42-1/4	2 x 2 x 3/16	40-1/2	9/16	36	24.50	
40	40-1/4	44-1/4	2 x 2 x 3/16	42-1/2	9/16	36	26.50	
42	42-1/4	46-1/4	2 x 2 x 3/16	44-1/2	9/16	40	27.50	
44	44-1/4	48-1/4	2 x 2 x 3/16	46-1/2	9/16	40	29.00	$44.84
46	46-1/4	50-1/4	2 x 2 x 3/16	48-1/2	9/16	44	30.00	
48	48-1/4	52-1/4	2 x 2 x 3/16	50-1/2	9/16	44	31.50	

Example of Round Black Iron
Takeoff and Extension

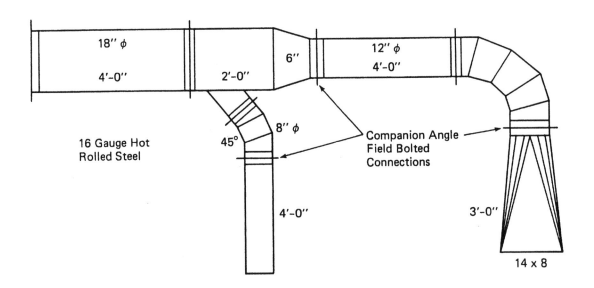

Per Piece Duct Takeoff Sheet

Job __Sand Blasting Exh.__ Mat'l __16 Ga.__ Duct Elev __12 FT__

Drwg/Flr __1st__ Sys. __E-1__ Pres __B.I.__ Lin/Insul _____

QTY	DUCT SIZE	TYPE DUCT	Equivalent LINEAL FEET Per Piece	TOT LF	WEIGHT LBS /LF	WEIGHT Total	SHOP Hrs /Pc	SHOP Total	FIELD Hrs /Pc	FIELD Total
1	18"φ	Pipe	4	4	14.1	56.4		1.1		2.3
1	"	45°RED TEE 4		4	14.1	56.4		6.1		2.3
1	12"φ	Pipe	4	4	9.4	37.6		.9		1.5
1	"	ELL 3		3	9.4	28.2		4.0		1.5
1	"	SQ to Rnd 3		3	9.4	28.2		4.8		1.5
1	8"φ	45° ELL	1	1	6.3	6.3		1.5		1.2
1	"	Pipe	4	4	6.3	25.2		.7		1.2
	Angle Rings	(Purchase)					(Weld to duct)			
3	18"φ	1			10.5	31.5	.74	2.2		
6	12"φ	1111			6.0	36.0	.60	3.6		
4	8"φ	11			2.3	9.2	.55	2.2		
			Totals			315 LB		27.1 hr		11.5 hr

Aluminum

Uses

Shower exhausts
Locker exhausts
Pool exhausts

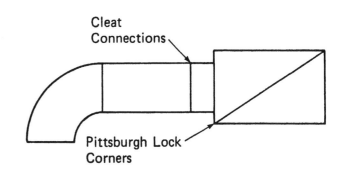

Limitations and Problems

Structurally weak
Low pressure systems only
Aluminum to aluminum doesn't slide well
800 degree maximum temperature
Crackage, slittage, metal fatigue

Gauges

Use gauge numbers for heavier than LP galvanized. Aluminum gauges usually referred to in 1000ths of an inch. Galvanized weighs about 3-1/4 as much as aluminum.

Example: 20x10 LP galv. weighs 7 lbs/LF w/20% aluminum
           20x10 2.32 lbs/LF

Galv LF	ALUMINUM			
Gauges	Gauge	Thickness	lbs/SF	Duct Size Range
26	22	.025	.353	0-12
24	20	.032	.452	13-30
22	18	.040	.564	31-54
20	16	.050	.706	55-84
18	14	.063	.889	85 up

<u>16 gauge or heavier</u>
<u>can be welded</u>

Types of Aluminum Use 3003 or 1100, soft or * hard.

Connections

Start bar cleats and reinforcing angles at 18" wide instead of 24" "S" and bar cleats can be used; metal flanges; and companion angles.

Reinforcing

Need double the amount of reinforcing angles that galvanized LP does, both in intermediate locations and at connections.

Budget Pricing

Typical exhaust system, .032" thick, 25% fittings by weight. Standard height run. Labor $39.00/hour, Materials $.65/lb, 20% allowance for waste, cleats, hangers, seams and hardware.
    $12.90/lb $5.83/sq ft

Labor Productivity Rates

In square feet per hour: 33 sq ft/hr, Field 19 sq ft/hr
In pounds per hour: 12 lb/hr, Field 7 lb/hr

Rectangular Heavy Gauge Ductwork
14 Gauge Aluminum, Angle Flange Connection
Shop Hours
Fittings Hours Per Piece; Straight Hours Per Foot

Manual Fabrication

Size	Straight	90 Elbw	45 Elbw	Sq/Elbw Trans	Tap	Wye	Sq/Rd	Offset	Door
12x12	1.25	1.93	1.44	1.44	1.14	1.70	3.63	2.17	1.53
18x12	1.44	2.31	1.74	1.68	1.34	2.01	4.03	2.52	1.60
24x12	1.64	2.69	2.02	1.92	1.55	2.32	4.42	2.87	1.67
30x12	1.82	3.07	2.31	2.15	1.75	2.63	4.82	3.23	1.74
36x12	2.01	3.46	2.59	2.39	1.95	2.93	5.25	3.58	1.81
36x18	2.20	3.84	2.88	2.62	2.15	3.25	5.76	3.93	1.88
42x18	2.39	4.84	3.63	2.86	2.35	3.56	6.26	4.28	.1.95
48x18	2.58	—5.28	3.96	3.09	2.55	3.86	6.76	4.64	2.02
54x18	2.76	5.72	4.30	3.33	2.76	4.17	7.26	4.99	2.09
60x18	2.95	6.17	4.63	3.56	2.96	4.48	7.77	5.34	2.16
72x18	3.33	9.23	6.93	4.03	3.36	5.10	8.78	6.05	2.30
84x18	3.70	10.41	7.81	4.50	3.76	5.72	9.78	6.75	2.44

Automated Fabrication

Size	Straight	90 Elbw	45 Elbw	Sq/Elbw Trans	Tap	Wye	Sq/Rd	Offset	Door
12x12	1.68	2.59	1.94	1.94	1.53	2.28	4.87	2.90	2.05
18x12	1.94	3.10	2.33	2.25	1.80	2.69	5.40	3.38	2.15
24x12	2.19	3.61	2.71	2.57	2.07	3.11	5.93	3.85	2.24
30x12	2.44	4.12	3.09	2.88	2.34	3.52	6.46	4.32	2.33
36x12	2.69	4.63	3.47	3.20	2.61	3.93	7.04	4.79	2.47
36x18	2.94	5.14	3.86	3.51	2.88	4.35	7.71	5.27	2.52
42x18	3.20	6.48	4.86	3.83	3.15	4.76	8.39	5.74	2.62
48x18	3.45	7.07	5.31	4.14	3.42	5.18	9.06	6.21	2.71
54x18	3.70	7.67	5.75	4.46	3.69	5.59	9.73	6.68	2.81
60x18	3.95	8.27	6.20	4.77	3.96	6.00	10.40	7.16	2.90
72x18	4.46	12.37	9.28	5.40	4.50	6.83	11.75	8.10	3.08
84x18	4.96	13.94	10.46	6.03	5.04	7.66	13.10	9.05	3.27

Rectangular Heavy Gauge Ductwork
14 Gauge Aluminum, Angle Flange Connection
Field Hours
Fittings Hours Per Piece; Straight Hours Per Foot

Size	Straight	45 Elbw 90 Elbw Wye	Sq/Elbw Trans Offset	Tap	Sq/Rd
12x12	1.20	1.08	1.20	0.29	0.87
18x12	1.43	1.28	1.43	0.35	1.01
24x12	1.65	1.49	1.65	0.40	1.16
30x12	1.88	1.69	1.88	0.45	1.31
36x12	2.10	1.89	2.10	0.51	1.45
36x18	2.33	2.09	2.33	0.56	1.59
42x18	2.55	2.30	2.55	0.62	1.73
48x18	2.78	2.50	2.78	0.67	1.88
54x18	3.00	2.70	3.00	0.72	2.02
60x18	3.23	2.90	3.23	0.78	2.16
72x18	3.68	3.31	3.68	0.89	2.45
84x18	4.13	3.71	4.13	0.99	2.74

STAINLESS STEEL

Uses

Utilized in laboratories, kitchens, plants, schools, for fume exhausts, dishwashers, pool exhausts, showers, locker rooms, ovens, open face tanks, etc.

Advantages

Resists chemical corrosion excellently, doesn't rust or stain, and withstands erosion very well.

Disadvantages

It is more difficult to shear, cut drill punch, to work with in general due to it's hardness. Dulls tools faster; heavier capacity machinery needed. Surface has to be protected from scratches and mars in processing if exposed. It is heavier than other corrosion resistant ductwork such as PVC or FRP. Material cost is far higher than regular steel.

Material Costs

Runs from $1 to $2 per pound depending on the quantity purchased, finish required, sheet sizes, gauge and if protective covering is needed. Average $2.60/lb.

Material Calculations

It's best to figure the actual sheet sizes needed rather than the weight per linear foot with a more or less arbitrary waste factor, because of the high cost.

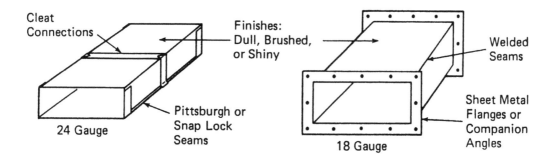

Example Cost Breakdown of Typical Duct Runs—with labor productivity rates.

Item	24 Gauge (.987 lb/sq ft) Cleats & Pittsburghs Sq Ft/Hr	Cost/Sq Ft	18 Gauge (2.016 lb/sq ft) Welded Seams, SM Flanges Sq Ft/Hr	Cost/Sq Ft
Material		$ 2.76		$5.30
Shop	30	1.33	10	4.00
Field	17	2.28	15	2.66
Drawings	145	0.28	145	0.29
Cartage	300	0.10		0.06
Total Direct Costs	6.75			12.37
Mark Up, 35%	2.34			4.32

Budget Figures,   sell $9.10 sq ft $16.68 sq ft
Labor based on $39.00/hour

Rectangular Heavy Gauge Ductwork
14 Gauge Stainless Steel, Angle Flange Connection—Shop Hours
Fittings Hours Per Piece; Straight Hours Per Foot

Manual Fabrication

Size	Straight	90 Elbw	45 Elbw	Sq/Elbw Trans	Tap	Wye	Sq/Rd	Offset	Door
12x12	2.80	4.31	3.23	3.23	2.55	3.80	8.11	4.84	3.41
18x12	3.23	5.16	3.88	3.75	3.00	4.49	9.00	5.63	3.58
24x12	3.65	3.01	4.51	4.28	3.45	5.18	9.88	6.41	3.74
30x12	4.06	6.86	5.15	4.80	3.90	5.86	10.76	7.20	3.89
36x12	4.49	7.71	5.79	5.33	4.35	6.55	11.73	7.99	4.05
36x18	4.90	8.56	6.43	5.85	4.80	7.25	12.85	8.78	4.20
42x18	5.33	10.80	8.10	6.38	5.25	7.94	13.98	9.56	4.36
48x18	5.75	11.79	8.85	6.90	5.70	8.63	15.10	10.35	4.51
54x18	6.16	12.78	9.59	7.43	6.15	9.31	16.21	11.14	4.68
60x18	6.59	13.78	10.34	7.95	6.60	10.00	17.34	11.93	4.83
72x18	7.43	20.61	15.46	9.00	7.50	11.39	19.59	13.50	5.14
84x18	8.26	23.24	17.44	10.05	8.40	12.76	21.83	15.08	5.45

Automated Fabrication

Size	Straight	90 Elbw	45 Elbw	Sq/Elbw Trans	Tap	Wye	Sq/Rd	Offset	Door
12x12	2.24	3.45	2.58	2.58	2.04	3.04	6.49	3.87	2.73
18x12	2.58	4.13	3.10	3.00	2.40	3.59	7.20	4.50	2.86
24x12	2.92	4.81	3.61	3.42	2.76	4.14	7.90	5.13	2.99
30x12	3.25	5.49	4.12	3.84	3.12	4.69	8.61	5.76	3.11
36x12	3.59	6.17	4.63	4.26	3.48	5.24	9.38	6.39	3.24
36x18	3.92	6.85	5.14	4.68	3.84	5.80	10.28	7.02	3.36
42x18	4.26	8.64	6.48	5.10	4.20	6.35	11.18	7.65	3.49
48x18	4.60	9.43	7.08	5.52	4.56	6.90	12.08	8.28	3.61
54x18	4.93	10.22	7.67	5.94	4.92	7.45	12.97	8.91	3.74
60x18	5.27	11.02	8.27	6.36	5.28	8.00	13.87	9.54	3.86
72x18	5.94	16.49	12.37	7.20	6.00	9.11	15.67	10.80	4.11
84x18	2.61	18.59	13.95	8.04	6.72	10.21	17.46	12.06	4.36

Rectangular Heavy Gauge Ductwork
14 Gauge Stainless Steel, Angle Flange Connection
Field Hours
Fittings Hours Per Piece; Straight Hours Per Foot

Size	Straight	45 Elbw 90 Elbw Wye	Sq/Elbw Trans Offset	Tap	Sq/Rd
12x12	2.00	1.80	2.00	0.49	1.45
18x12	2.38	2.14	2.38	0.58	1.69
24x12	2.75	2.48	2.75	0.66	1.94
30x12	3.13	2.81	3.13	0.75	2.18
36x12	3.50	3.15	3.50	0.85	2.41
36x18	3.88	3.49	3.88	0.94	2.65
42x18	4.25	3.83	4.25	1.03	2.89
48x18	4.63	4.16	4.63	1.11	3.13
54x18	5.00	4.50	5.00	1.20	3.36
60x18	5.38	4.84	5.38	1.30	3.60
72x18	6.13	5.51	6.13	1.48	4.08
84x18	6.88	6.19	6.88	1.65	4.56

FRP—FIBERGLASS REINFORCED PLASTIC DUCTWORK

Uses

Primarily used in corrosive exhaust systems for chemicals, acids, caustics, etc.

Advantages

It is relatively light weight in comparison to steel, highly corrosive resistant, structurally stronger than plastic ductwork, can be worked in a wider range of temperatures, can be molded into many shapes and seals almost perfectly.

Description

FRP is a combination of 1/16" thick laminated layers of fiberglass mesh and epoxy and resin. The plastic is sprayed onto wood molds in the various duct configurations. Then stranded fiberglass mesh is laid over it which in turn is covered with spray. This process goes on until the desired thickness is attained.

Connections

Connections can be flanged, slip over, coupling or butt type and are chemically field welded.

Materials run 1/8", 3/16", 1/4" thick depending on the duct size and application, and cost an average of $3.00 per square foot.

Budget costs installed, for the typical mix and average size, is approximately $16.05/sf This is based on $39.00/hr labor, $3.86/sq ft for material and a 35% markup for overhead.

Fabrication is highly specialized with the chemical process, molds, expertise and it is normally better to purchase the ductwork from a FRP manufacturer. It runs about S 10. 50 per square foot depending on the mix of fittings and straight duct, and on the average size.

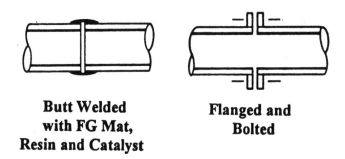

Butt Welded with FG Mat, Resin and Catalyst

Flanged and Bolted

PVC DUCTWORK

Uses

Used in corrosive exhaust systems, fume, chemical, mist, etc. and is resistant to many chemicals.

Advantages

PVC is lightweight, it can be bought in sheets, and it can be easily cut, formed, heat welded or joined with a solvent.

Description

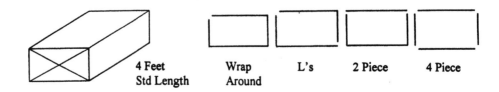

4 Feet   Wrap   L's   2 Piece   4 Piece
Std Length   Around

Materials

PVC is a poly vinyl chloride, smooth surface, thermo plastic. Usually 1/8" thick sheets are used for smaller ducts and 3/16" thick for larger ones. Materials cost about $3.86/sf

Connections

Connections are generally, either flanged or coupling.

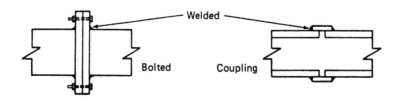

Fabrication Sheets

Fabrication sheets can be cut in a shear or with a saw. There are four methods for bending material; you can put the sheet in a long flat oven and when heated to the proper temperature, remove it and bend it. A second method is with an electric calrod which is laid on the sheet on the bend line, the line is heated and then bent in a hand brake. The third method is with a torch and the fourth with a heat table and asbestos.

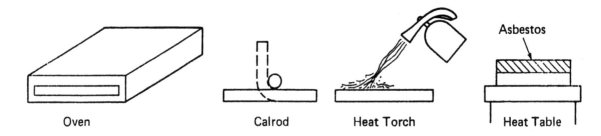

Oven   Calrod   Heat Torch   Heat Table

Seams and Connections

Seams and connections can be chemically welded with a solvent that fuses the material together, heat welded with a torch, or with a wire feed welding gun that runs at 40" per minute.

Shop Labor runs about 23 sf/hr for typical mixes of fittings, pipe and sizes.

Field Labor productivity rate average 16 sq ft/hr.

Budget Figures typical mix, average size $13.10/sq ft.

PVC COATED GALVANIZED

Uses

Used either in corrosive exhaust systems for fumes, chemicals, acids, caustics or for underground HVAC supply systems.

Description

PVC coated galvanized, commonly called by its registered trade name "PVS," is galvanized material with a protective coat of corrosion resistant PVC on it. It comes in either sheets of rectangular ductwork, spiral pipe and fittings, or in residential lock seam pipe and fittings. Usually the corrosive side has a four mil thickness on it and the non corrosive side a one mil thickness.

Construction and Connections

Rectangular ductwork fabricated from sheets can be seamed with pittsburghs, bent, sheared, etc. the same as regular galvanized. Where the PVC coating may be scratched or marred it can be touched up. Connections can be "S" and drive cleats or flanged metal. Either PVC tape and cement or hardcast tape can be used to seal connections.

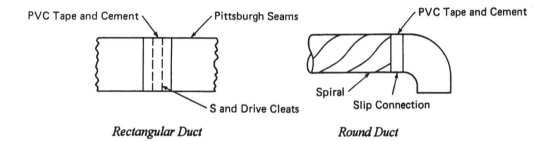

Rectangular Duct   *Round Duct*

Per Piece Labor FRP, PVC and Aluminum
Flanged Connections
Straight Duct and Fittings Mixed

Size	Semi Perim	FRP		PVC		Aluminum 16 GA	
		Shop	Field	Shop	Field	Shop	Field
12 x 6	18	1.9	1.4	0.9	1.3	1.8	1.3
12 x 12	24	2.0	1.5	1.0	1.4	1.9	1.4
18 x 12	30	2.3	1.7	1.2	1.6	2.1	1.5
18 x 18	36	2.5	1.8	1.3	1.7	2.3	1.6
24 x 12	36	2.6	1.9	1.3	1.8	2.4	1.7
24 x 18	42	2.8	2.1	1.4	2.0	2.6	1.9
30 x 12	42	2.9	2.3	1.5	2.2	2.7	2.0
30 x 18	48	3.1	2.5	1.6	2.3	2.9	2.2
36 x 12	48	3.2	2.6	1.7	2.4	3.0	2.3
36 x 18	54	3.6	2.9	1.9	2.7	3.3	2.6
36 x 24	60	3.8	3.0	2.0	2.8	3.5	2.7
42 x 18	60	3.9	3.0	2.0	2.9	3.6	2.7
48 x 18	66	4.3	3.2	2.3	3.0	4.0	2.8
48 x 24	72	4.6	3.3	2.4	3.1	4.2	2.9
54 x 18	72	4.8	3.4	2.5	3.2	4.4	3.0
60 x 18	78	5.2	3.8	2.7	3.6	4.8	3.4
66 x 18	84	5.4	4.3	2.8	4.1	5.0	3.8
72 x 18	90	5.7	4.8	2.9	4.5	5.3	4.3
72 x 24	96	6.1	5.3	3.2	5.0	5.6	4.7
78 x 24	102	6.3	5.8	3.3	5.4	5.9	5.1
84 x 24	108	6.6	6.2	3.5	5.8	6.1	5.5

Material

PVC coated galvanized comes in various gauges 26, 24, 22 etc. The most commonly used gauges are 24 and 22. Typical cost is about $2.37 per square foot for 24 gauge sheets.

Labor

Fabrication labor runs an average of 30 sq ft/hr. Installation is typically 17 sq ft/hr.

Budget Costs

Typical budget installed cost for average 24 gauge system with 25% fittings by square feet runs $8.83/sq ft.

LABOR MULTIPLIERS
FOR HEAVY GAUGE DUCTWORK

The following labor multipliers should be applied for various conditions:

1. If two exact **DUPLICATE** fittings are being fabricated, reduce the time 33% on the second elbow because of the saving on the layout time. Both elbows can be further reduced 10% because of set up time savings on a production run approach.

2. Fabrication labor can be reduced 33%, if TEMPLATES are being used in the hand layout of the gores.

3. A fast working highly skilled sheet metal layout man can work 25% to 50% faster than the average mechanic.

To relate your own productivity rates to the Super-Duct program you must first of all adjust the per piece labor factors on an overall basis for your company, and then adjust factors for individual job conditions such as duplications, templates, etc.

Also, duct specs must be set up properly for plasma cutters, stick or wire feed welding, continuous or tack weld, type welding machinery used, etc. before making takeoff so that proper labor factors are activated.

SAMPLE LABOR TESTS
ON ROUND BLACK IRON ELBOWS

The following table shows the fabrication labor variations for a 48" dia, 90 degree, 5 gore elbow with angle rings welded on when done by hand, on a plasma cutter or with a gore cutting machine:

	Shop		Field	
	10 ga	14 ga	10 ga	14 ga
Hand Layout and Cut	21.14	17.62	7.94	5.47
With Plasma Cutter	15.86	13.72	7.94	5.47
With Gore Machine	13.96	11.63	7.94	5.47

The hours calculated are based on the custom fabrication of a single item, with individual layouts and separate setups of machinery. No templates are being used in the hand fabricated sample.

AIR POLLUTION ESTIMATING

Air pollution control and industrial exhaust ventilation work today comprise a major industry of more than $2 billion annually.

Among the factors that have contributed to the strength of this industry are social concern over the quality of air both inside and outside environments; OSHA and EPA monitoring and policing; increased realization of the actual damages of uncontrolled pollutants, including black lung, cancer, blindness, crippling, and death; and the many lawsuits filed by those injured by uncontrolled contaminants.

A large portion of air pollution control work occurs in processing plants of major industries such as chemical, coal mining, pharmaceutical, woodworking, etc., followed by manufacturing plants, laboratories, restaurants, auto repair garages, etc.

Items and operations that require exhausting include boilers, ovens, kilns, dryers, welding,' soldering, painting, and metal cleaning and treating. Additional operations include waste conveying, milling, spraying, cooling, raw product handling, crushing, grinding, screening, grating, bagging and packing, polishing, buffing, melting, burning, and shakeout.

The costs of air pollution control systems can vary from 11¢ per cfm for a simple propeller fan installed in a wall or window to exhaust smoke to $6 to $7 per cfm for a complex open tank exhaust system including makeup air, special ducts, hoods, etc.

For contractors engaged in design and construction, these installations may carry with them a risk in terms of performance. If a system is improperly designed, or if attempts to keep costs down result in an inadequate system, the consequences can be costly. A $5,000 collector may have to be replaced with a $10,000 unit or hoods remade at a contractor's expense. To estimate costs completely and accurately, the contractor must be knowledgeable of the basic engineering principles and considerations involved in properly designing workable systems.

The Purpose of Control

The purpose of an air pollution control system is to:

1) Remove harmful contaminants such as dusts, fumes, mists, vapors, gases, and heat from a source area.

2) Transport them through ducts to a collector or fan.

3) Separate the contaminants from the air and either reclaim or dispose of them, or discharge the mixture directly to the atmosphere if permissible.

ENGINEERING FUNDAMENTALS

Contaminants include anything that is harmful to humans, animals, plants, or property in the internal or external environment. There are two basic categories of contaminants, particulates and gases.

Particulates are small solid or liquid particles such as dusts, powders, smoke, liquid droplets, and mists. Fly-ash from coal furnaces and asbestos dusts are examples of harmful particulate pollutants. Gas pollutants are fluids without form that occupy space rather uniformity, such as carbon monoxide or chloroform. A fume is an irritating smoke, vapor, or gas.

Contaminants are measured in microns. A micron is smaller than the point of a sharpened pencil. There are 10,000 microns in a centimeter, 25,400 in 1 inch. Particles smaller than 2 microns are visible only with an electronic microscope, those between 2 and 10 microns with a standard microscope, and those larger than 10 microns with the naked eye.

Pollen and lint range in size from approximately 5 to 70 microns; smoke from approximately 0.001 to 1 micron. averaging approximately 0.1, and 90 percent of atmospheric dusts are approximately 0.25 microns in size, but they can range up to 30 or 40 microns.

An important consideration in air pollution control is the **harmful effects of the contaminants** involved. Pollutants can be *toxic* (poisonous, possibly fatal) or *noxious* (harmful to health but not necessarily deadly). They may be *corrosive* (alkalis, acids, salts, and caustics) or *erosive* (causing wearing away by abrasion). Processes can be *heat* producing, *moisture* producing, or *smoke* producing, causing uncomfortable or harmful levels. *Odors* my be offensive or harmful. Pollutants can be *inflammable, explosive,* or *radioactive.* And finally, contaminants include *bacteria* and *viruses.*

Fire and explosion hazards are critical factors. Dusts as well as many fumes and mists have flash point temperatures, at which spontaneous combustion occurs, and explosive limits, which are percentage ranges defining when explosion occurs.

There are **two flash points** for a hazardous pollutant, one for a closed situation and the other for an open situation. There are also two explosive limits, a lower and an upper. For example, ethyl acetate has flash points of 24°F and 30°F and lower and upper explosive limits of 2.18 and 11.4 percent. Similarly, methanol has flash points of 54°F and 60°F and explosion limits of 6.72 and 36.5 percent.

With regard to toxicity, the point at which a mixture of pollutant and air starts to become toxic is called it's **threshold limit** value. There are also different levels of acceptable concentrations for different lengths of exposure. Threshold limits and prolonged exposure limits are usually stated in terms of ppm, parts of contaminants per million parts of air, at a certain temperature and pressure. For example, ammonia has a threshold limit value of 25 ppm, a daily exposure limit of 100 ppm, and a 1/2 to 1 hour exposure limit of 2500 ppm; it is rapidly fatal at 5000 ppm.

Other measures of contaminant concentration or density sometimes used in pollution control work includes weight per cubic foot, pounds per hour, grains of moisture-per pound of air, percentage of gas, and others.

Another important factor affecting the performance of a system is the **settling rate of a solid** or liquid particles. This is a measure of the time it takes for a particle to float down and come to rest, and is dependent on particle size. A 50 micron particle will free fall at 14.8 fpm whereas a 1 micron particle will free fall at 0.007 fpm. Thus, the former will fall almost 15 ft in 1 minute; the latter will take approximately 36 hours to fall the same distance.

AIR POLLUTION SYSTEM DESIGN

The general procedure in system design is to identify the problem, analyze the problem, establish facts, determine the magnitude of the pollution problem, calculate capacities, design hoods and ductwork, select system components (collectors, fans, makeup air units, etc.), and choose an appropriate disposal method. (See figure on following page.)

Some of the important considerations in this area are:

Capture Velocity

One of the major considerations involved in the design of an air pollution control system is the velocity required to pick up the contaminants from their source area. This velocity is called the capture velocity, and it

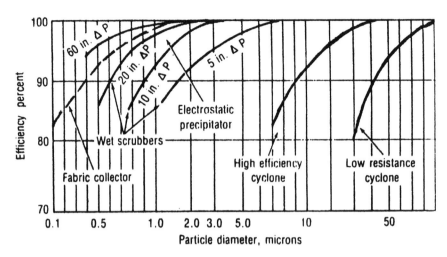

Collector efficiency ranges

must overcome spillage, air and heat currents, and the contaminants' weight and velocity. For example, laboratory fumes heavier than air may require a capture velocity at their point of emission of 50 fpm to overcome their tendency to spill to the floor as well as any currents of air flowing across the source area. Sawdust, which weighs approximately 12 lb/cuft, usually requires a suction pressure of 2.5 in. WG for proper pickup.

Transport Velocity

The air velocity required in the duct to carry the material, overcome resistance and turns, lift it in risers, and convey it to a collector or fan is the transport velocity. Powdered coal requires a transport velocity of approximately 4000 fpm.

Hood Face Velocity

This is the velocity required at the face of a hood for proper contaminant pickup and suction into the hood. A typical slot hood face velocity might be 2000 fpm; a kitchen exhaust hood might require a face velocity of only 100 to 150 fpm. Hoods must be designed to draw in pollutants properly. Pollutants can be captured by completely enclosing the source area, by providing a local hood at the exact point of emission, or by providing a general exhaust hood or grille. There are an infinite number of hood configurations, and the best designs are those in which contaminants project directly into the hoods. For example, particulates from grinding, buffing, polishing, or saw wheels should be propelled directly into their hoods. For heat removal, fumes should rise into their hoods.

Air Volume

Factors that determine the volume of air required are the surface area of the source, the quantity of material

being removed, the amount of dilution required, the size of hood opening, and the face velocity. A kitchen hood with a 30 sq ft opening and a 100 fpm face velocity requires 3000 cfm.

Ductwork

There are two general approaches to designing a duct system. One is the balanced static regain approach, in which the system is automatically balanced through exact pressure design of the ductwork This can be done when all hoods are used whenever the system is in operation. The second basic approach involves blast gates. Dampers are put in branches to control air flow, and the system is not designed to be automatically balanced. Here, velocities are the main consideration. Total system static pressure is contingent on the run with the greatest resistance as well as on the pressure drops of the collector and other components.

Velocity

Some systems are designed for 100 percent usage of hoods all the time, others for only partial use of hoods at any one time. In sizing collectors and fans, the diversity factor must be taken into account to determine whether they should handle 80, 60, or some other percentage of the design air volume and static pressure. If only 60 percent of a system is in operation at any given time, it is very costly to size for larger capacities.

Material removal rate

Another factor that must be integrated into the design is the removal rate for dusts, particles, etc. The density of the material, its concentration in the air stream, the quantity of air involved, and the quantity of material to be removed must be considered to arrive at an ultimate removal rate.

References

Industrial Ventilation, A Manual of Recommended Practice, 15th ed., American Conference of Governmental, Industrial Hygienists, Committee on Industrial Ventilation, Lansing, Michigan, 1978.

Alden, John, and Kane, John, *Design of Industrial Exhaust Systems*, 4th ed., Industrial Press, Inc., New York, 1970.

ASHRAE Handbook & Product Directory, 1976 systems, Chapters 21-23, American Society of Heating, Refrigerating and Air-conditioning Engineers, Inc., New York.

Wendes, Herbert C., *Everything You Ever Wanted to Know About Sheet Metal Estimating*, Wendes Systems, Inc., Arlington Heights, Illinois, 1976.

COST ESTIMATING PROCEDURES

In general, the cost estimating procedure is as follows:

1. If the system is not yet designed, the first step is to develop an effective and economic design that will perform properly and meet codes.

2. If the system is already designed, a thorough study of the plans and specifications is required so that the estimator will be familiar with all of the requirements and the scope of the project.

3. Equipment suppliers and subcontractors should be notified.

4. The estimator makes his/her takeoff of ductwork, sheet metal specialties, and equipment.

5. Calculations are run on ductwork, specialty material and labor.

6. All items are transferred to a summary sheet. Any pricing not yet done is filled in, labor is extended, and all purchases and labor are to be totaled.

7. Markups, taxes, permits, increases, etc., are applied in a recap.

Air Pollution Estimate
Scope and Check List

Scope and checklist for air pollution control estimates. Many items must be included in an industrial exhaust or air pollution control estimate, and some of them are easily missed. Great care must therefore be taken to make sure they are all accounted for. This table provides a checklist but is not necessarily all inclusive.

EQUIPMENT
- ❏ Collectors
- ❏ Platforms
- ❏ Ladders
- ❏ Stands
- ❏ Fans
- ❏ Screw Conveyors
- ❏ Dampers
- ❏ Makeup Air Equipment
- ❏ Heat Recovery Equipment
- ❏ Blast Gates
- ❏ Starters
- ❏ Pumps
- ❏ Compressors
- ❏ Vibration Isolators
- ❏ Inlet Vane Dampers
- ❏ Louvers
- ❏ Shutters

DUCTWORK
- ❏ Black Iron
- ❏ Stainless Steel
- ❏ Aluminum
- ❏ Galvanized
- ❏ Aluminized Steel
- ❏ Corten
- ❏ PVC
- ❏ PVC Coated Galvanized
- ❏ FRP
- ❏ Flues
- ❏ Flexible Tubing

SHEET METAL SPECIALTIES
- ❏ Hoods
- ❏ Dampers
- ❏ Turning Vanes
- ❏ Hoppers
- ❏ Platforms
- ❏ Flex Connections
- ❏ Access Doors
- ❏ Walk Thru Doors
- ❏ Gaskets
- ❏ Sealants
- ❏ Painting Ducts

SPECIAL LABOR
- ❏ Shop Drawings
- ❏ Field Measuring
- ❏ Cartage
- ❏ Testing and Balancing
- ❏ Startup
- ❏ Service

SUBS, RENTALS
- ❏ Gas Piping
- ❏ Water Piping
- ❏ Pneumatic Piping
- ❏ Electrical
- ❏ Temperature Controls
- ❏ Concrete Pads
- ❏ Crane
- ❏ Cutting and Patching

Smyth Lighting Fixture Plant
Sample Estimate

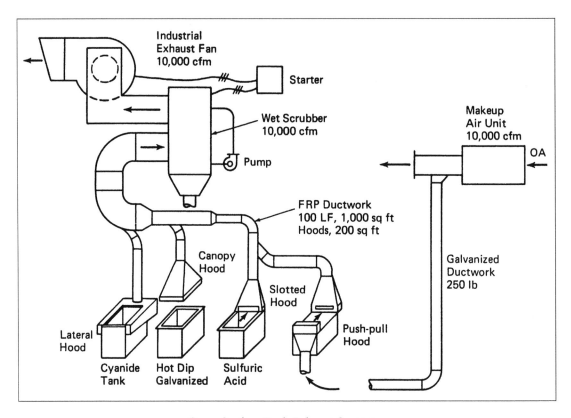

Open Surface Tank Exhaust System

Estimate Summary and Extension Sheet

Job __SMYTH LIGHTING__ __FIXTURE PLANT__

Open Surface Tank Exhaust Estimate

		MATERIAL COST		SHOP		FIELD	
		UNIT	TOTAL	UNIT	TOTAL	UNIT	TOTAL
1	Wet Scrubber, 10,000 CFM	# 1.97/CFM	$19,700	—			45
1	Pump		700	—			—
1	Makeup Air Unit, Direct Fired	.82/CFM	8,177	—			24
1	Industrial Fan, 10,000 CFM, 25 HP	.66/CFM	6,567	—			21
2	Starters		1,700	—		2	4
	FRP Ductwork, 1000 SF		3,000	12 SF/HR	83	15 SF/HR	67
	Galv. Ductwork, 250 LBS	.41/LB	104	40 LBS/HR	7	50 LBS/HR	13
4	Hoods, FRP, 200 SF		600	7	28	4	16
	Field Measure				—		8
	Testing and Balancing; Svc.				—		4
	Cartage, 4 Loads			3	12		—
	Total Eqpt. Mat.		$40,548	Totals	130 Hrs.		202 Hrs.
Subs	Wiring 150ft, 54 amps	# 10.33/LF	1,551				
	Crane 1/2 day		400				
	Water Piping Valves		700				
	Total Subs		$2,497				
	RECAP						
•	Equipment, Materials		$ 40,548				
	Markup on Eqpt. 30% Ovhd.		12,164				
•	Labor, 332 Hrs. @ $33.00		10,956				
	Markup on Labor, 35% Ovhd.		3,267				
•	Subs		2,497				
	Markup on Subs, 10%		249				
•	Total Costs with Ovhd. Mu		$ 69,681				
	Profit 5%		3,403				
	Sales Tax on Mat., 7%		2,838				
•	Total Sell Price		$ 75,922				

This page intentionally left blank

Chapter 14

Sheet Metal Specialties and Acoustical Lining

This chapter covers estimating sheet metal specialties, acoustical lining for ductwork and other fabricated items, but are not ductwork per se, and are as follows:

- Duct Turning Vanes for Ductwork
- Splitter Dampers
- Flexible Connectors to Equipment

- Belt Guards
- Exhaust Hoods
- Equipment Platforms
- Fabricated Roof Hoods
- Sheet Metal Housings
- Housing Access Doors
- Blank Offs in Housings
- Coil Drain Pans

Access Doors, Belt Guards, Drain Pans

		MATERIAL		LABOR HOURS		DIRECT MATL &
	SIZE	WEIGHT	COST	SHOP	FIELD	LABR COST
CASING ACCESS DOORS	36x20	38 Lb	$43	3	3	$277
	48x20	50 Lb	$54	4	3	$327
	60x20	70 Lb	$60	4	3	$333
BELT GUARDS	3 ft	80 Lb	$54	3	2	$249
	5 ft	130 Lb	$83	4	3	$356
	7 ft	180 Lb	$124	5	4	$475
	10 ft	260 Lb	$182	6	5	$611
ANGLE IRON	5 ft	125 Lb	$117	8	3	$546
	7 ft	175 Lb	$155	9	4	$662
DRAIN PANS UNDER COILS	10' x 4' Galv	110 Lb	$78	3	1	$234
	10' x 4' S.S.	110 Lb	$399	4	3	$672
DRIP THROUGHS	10' x 1' Galv	30 Lb	$25	2	2	$181
	10 x 1' S.S.	30 Lb	$108	3	2	$303

Flexible Connections, Hoods, Stands and Platforms

		MATERIAL		LABOR HOURS		DIRECT MATL &
	SIZE	WEIGHT	COST	SHOP	FIELD	LABR COST
FLEXIBLE CONNECTIONS	18x12	5 ft	$4	0.8	0.7	$63
	36x18	9 ft	$8	0.9	1.6	$105
	48x30	13 ft	$11	1	1.8	$120
	60x42	17 ft	$14	1.1	2	$135
	84x60	24 ft	$20	1	3.2	$184
	108x72	30 ft	$25	1	4.1	$224
		(Perim)				
KITCHEN HOODS	16'x4'x2'	570 Lb	$263	30	24	$4,967
	(2 Sect)	4 Lites		Fire Protection		
		8 "V's"	$928	System		
		16 Filters	$875			$3,521
	8'x3'x2'	230 Lb	$667	16	12	$2,592
	5'x2'x2'	150 Lb		10	8	$1,296
FUME HOODS - WELDING HOODS	4'x3'x1.5' Galv Pittsburgh	72 Lb	$35	3	3	$269
EQUIPMENT OR COIL STANDS	10' x 2'x1'	55 Lb	$37	4	4	$349
	10' x 4'x2'	75 Lb	$52	5	4	$403
FAN PLATFORM	36x36	18 Lb	$25	2.5	3	$239
	48x48	25 Lb	$35	3	4	$308
	60x60	35 Lb	$47	4	5	$398

Diagram labels for KITCHEN HOODS: 16GA BI NFPA-96; Vapor Proof Lights $100 ea.; Continuous Welding External; Grease Filters; Alt. Lites; "V" Bank Assembly $47 ea.; Island Hood 150 cfm/sf; Wall Hood 100 cfm/sf

Diagram labels for FLEXIBLE CONNECTIONS: Metal Edge; Band Iron; Fabric

Diagram labels for FUME HOODS: 20 Ga Galv. or S.S.; Weld, Pittsburgh or Rivit Seams

Diagram labels for EQUIPMENT OR COIL STANDS: Welded Angle Stand 2" x 2" x $\frac{3}{16}$" $\angle$ S

Diagram labels for FAN PLATFORM: 1" Plywood; Angles; 14 Ga Formed Metal (s + $1\frac{1}{2}$ hr)

Roof Hoods

	SIZE	MATERIAL		LABOR HOURS		DIRECT MATL & LABR COST
		WEIGHT	COST	SHOP	FIELD	
FLEXIBLE CONNECTIONS	18x12	5 ft	$4	0.8	0.7	$63
	36x18	9 ft	$8	0.9	1.6	$105
	48x30	13 ft	$11	1	1.8	$120
	60x42	17 ft	$14	1.1	2	$135
	84x60	24 ft	$20	1	3.2	$184
	108x72	30 ft	$25	1	4.1	$224
		(Perim)				
KITCHEN HOODS	16'x4'x2'	570 Lb	$263	30	24	$4,967
	(2 Sect)	4 Lites		Fire Protection		
		8 "V's"	$928	System		
		16 Filters	$875			$3,521
	8'x3'x2'	230 Lb	$667	16	12	$2,592
	5'x2'x2'	150 Lb		10	8	$1,296
FUME HOODS - WELDING HOODS	4'x3'x1.5' Galv Pittsburgh	72 Lb	$35	3	3	$269
EQUIPMENT OR COIL STANDS	10' x 2'x1'	55 Lb	$37	4	4	$349
	10' x 4'x2'	75 Lb	$52	5	4	$403
FAN PLATFORM	36x36	18 Lb	$25	2.5	3	$239
	48x48	25 Lb	$35	3	4	$308
	60x60	35 Lb	$47	4	5	$398

FLEXIBLE CONNECTIONS — Metal Edge, Band Iron, Fabric

KITCHEN HOODS — 16GA BI NFPA-96, Vapor Proof Lights $100 ea., Continuous Welding External, Grease Filters, Alt. Lites, "V" Bank Assembly $47 ea., Island Hood 150 cfm/sf, Wall Hood 100 cfm/sf

FUME HOODS - WELDING HOODS — 20 Ga Galv. or S.S., Weld, Pittsburgh or Rivit Seams

EQUIPMENT OR COIL STANDS — Welded Angle Stand 2" x 2" x $\frac{3}{16}$" ∠ S

FAN PLATFORM — 1" Plywood, Angles, 14 Ga Formed Metal (s + $1\frac{1}{2}$ hr)

Turning Vanes

Weight
Two inch air foil weighs about 4.2 lbs/sq ft assembled
Duct width has to be multiplied by 1.4 trip factor to
come up with true length.

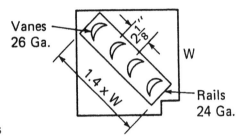

Costs to Purchase Formed Length
Approximately $0.74 per lb. in uncut, unassembled pieces, vanes and rails
(2″ vanes, $0.36/LF) (rails, $0.37/LF)

SIZE	SEMI-PERIM Inches	WEIGHT Lbs	SQ FT 45 Diag.	DIRECT MATERIAL COST	LABOR Man Hours	TOTAL MATERIAL & LABOR Direct Cost	With 30% O&P
12x6	18	3	0.7	$3.11	0.3	$14.81	$19.25
12x12	24	6	1.4	6.22	0.3	17.92	23.29
18x12	30	9	2.1	9.32	0.4	24.92	32.40
24x12	36	12	2.8	12.43	0.5	31.93	41.51
30x12	42	15	3.5	$15.54	0.6	$38.94	$50.62
30x24	54	29	7.0	30.04	0.8	61.24	79.62
36x12	48	18	4.2	18.65	0.8	49.85	64.80
36x18	54	26	6.3	26.94	0.8	58.14	75.58
36x24	60	35	8.4	36.26	0.9	71.36	92.77
42x18	54	31	7.4	$32.12	0.9	$67.22	$87.38
42x36	78	62	14.7	64.23	1.2	111.03	144.34
48x18	66	35	8.4	36.26	1.0	75.26	97.84
48x36	84	71	16.8	73.56	1.3	124.26	161.53
54x18	72	40	9.5	41.44	1.1	84.34	109.64
54x36	90	79	18.9	81.84	1.4	136.44	177.38
60x18	78	44	10.5	$45.58	1.2	$92.38	$120.10
60x36	96	88	21.0	91.17	1.5	149.67	194.57
72x18	90	53	12.6	54.91	1.5	113.41	147.43
72x36	108	106	25.2	109.82	1.8	180.02	234.02
72x54	126	159	37.8	164.72	2.6	266.12	345.96
84x18	102	62	14.7	64.23	1.8	134.43	174.76
84x36	120	123	29.4	127.43	2.0	205.43	267.06
84x54	138	185	44.1	191.66	3.0	308.66	. 401.26

Correction Factors on 2″ Air Foil	Material	Labor
1. Single skin vanes, 2″ radius	0.90	0.80
2. Air *foil* vanes, 4″ radius	2.00	0.80
3. Single skin vanes, 4″ radius	1.30	0.70

Labor hours include, time to cut, assemble and install.

Splitter Dampers

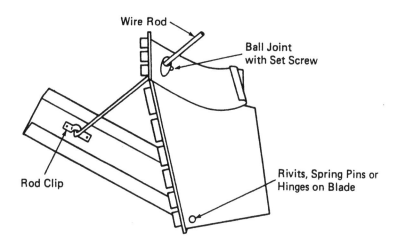

SIZE	SEMI-PERIM ——— Inches	WEIGHT Lbs	SQ FT 45 Diag.	DIRECT MATERIAL COST	LABOR ——— Man Hours	TOTAL MATERIAL & LABOR	
						Direct Cost	With 30% O&P
9x9	18	24	0.8	$5.27	0.3	$16.97	$22.06
18x12	30	22	2.6	6.22	0.3	17.92	23.29
30x16	46	20	6.7	8.41	0.5	27.91	36.28
42x18	60	20	9.0	9.61	0.5	29.11	37.84
54x20	74	20	13.0	12.98	0.8	42.23	54.90
66x24	88	is	29.0	22.01	0.8	51.26	66.63
78x30	108	18	43.0	30.06	1.0	69.06	89.78

Takeoff and Pricing Methods

1.   Determine approximate average size and multiply labor and material costs times total quantity.

2.   Add one linear foot of ductwork for each splitter damper to cover labor and material.

3.   Takeoff each individually and price in size groups.

SHEET METAL HOUSINGS

Built up sheet metal housings are field assembled casings used to enclose HVAC component equipment such as filters, coils, fans, water eliminators, dampers, etc.

Construction

Casings are built in panels with standing seams or channel flanges, 20", 26", 32" or 44" wide by lengths of anywhere between 5 and 10 feet and are normally 18 gauge. (See figure below.)

Material Calculations

1. Measure size of housing and calculate square footage.
2. Add 30% for waste, seams and hardware.
3. Measure required angles and add 15% waste.

A **typical full size panel** is 32" wide seam to seam, 8 feet long, requires an 18 gauge 36x96 sheet which weighs 52 pounds and costs $23.34 per sheet at $.42/lb.

Labor   Single Skin Panels

Fabrication	1/2 hr/panel	36 sf/hr	64 lbs/hr
Installation	2 hr/panel	9 sf/hr	20 lbs/hr

Includes angle and caulking labor.

Budget Figures

Single skin	$160/panel	$9.42/SF	$4.10/lb
Double skin	$314/panel	$17.41/SF	$8.71/lb

Labor and budget figures are based on an average 32" wide by 6 foot long panel, which is a typical size in a mix, single skin, 18 gauge galvanized, 39 lbs per panel, $39.00/hr, $.40/lb, 30% waste allowance and a 30% markup for overhead and profit.

Correction Factors

Double skin, 2" thick, 2" internal insulation, and perforated inner panel.

Material costs 3.0
Shop labor 3.0
Field labor 1.5

ACOUSTIC LINING

The main purpose of internal lining is to absorb sound, but it can also function simultaneously as a thermal insulator when needed. Sometimes it is simply used in place of insulation for economic or other reasons.

Uses

Lining is used in auditoriums, libraries, in outside ductwork, in high pressure ductwork off of a fan, on the low pressure side of high pressure terminal units.

Lining is not used in fume or partial exhaust industrial exhaust systems, in hot systems such as kitchen exhaust, in wet or moist situations such as with dishwashers, showers and pool exhausts.

Description

Fiberglass lining comes in 2 basic types blanket and rigid board. (See figure on following page.)

Calculating Material

1. If you have the square footages just add 15% for waste and corner overlaps.

2. If you only have poundage figures you have to convert them back to square feet by dividing the weight per square foot for each gauge into the poundage to get back to the square footage.

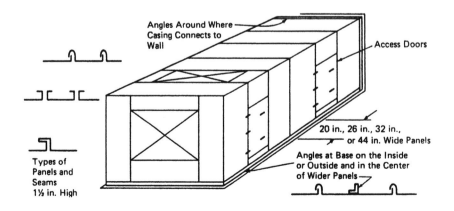

Types of Panels and Seams 1½ in. High

Angles Around Where Casing Connects to Wall

Access Doors

20 in., 26 in., 32 in., or 44 in. Wide Panels

Angles at Base on the Inside or Outside and in the Center of Wider Panels

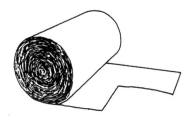

Blanket Rolls
4 ft Wide and 100 ft long
400 SF
Thicknesses: 1/2, 1, 1 1/2, & 2 inches
Densities: 1 1/2, 2 & 3 pounds

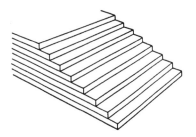

Rigid Board
Sizes: 2 ft x 4 ft, 3 ft x 6 ft
Thicknesses: 1 & 2 inches
Densities: 3, 4, 5 or 6 lbs

Application

1. Pins spot welded onto bare metal, lining pressed over pins, capped with washers.

2. Or "Grip Nails" are shot through the lining after it is applied.

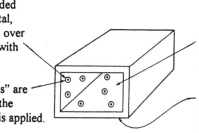

The air side face of the lining is either neoprene coated or an interwoven surface to prevent erosion.

Cement is sprayed, rolled or brushed on the bare metal first and then the lining is stuck on top.

Example:

26 ga 5,000 lbs	.906 lbs/sq ft =	5,519 sq ft
24 ga 15,000 lbs	1.156 lbs/sq ft =	12,519 sq ft
22 ga 8,000 lbs	1.406 lbs/sq ft =	5,690 sq ft
20 ga 4,000 lbs	1.656 lbs/sq ft =	2,415 sq ft
18 ga 2,000 lbs	2.156 lbs/sq ft =	928 sq ft

If the poundage figures already have the standard 20% waste in them, you them must reduce the square feet 5% to bring it down to 15%.

Material Cost

1" thick, 1-1/2 lb density	$0.43/sq ft
1/2" thick, 2 lb density	$0.35/sq ft
Cement and Pins, add	$0.10/sq ft

Increased Duct Size

Increase the metal duct size to cover the lining thickness. For example, increase a 20 x 10 duct to 22 x 12 for 1" thick lining. Weight increases about 12 percent for average duct sizes for 1 inch thick lining and 6 percent for 1/2 inch thick.

Labor

(Based on 1" thick, 1-1/2 lb density, blanket.)

Correction Factors on Labor

1. 1" thick, 3 lb 1.15
2. 2 inch thick, 1-1/2 lb 1.15

Square Feet Per Hour

Ratio Fitting SqFt to Total SqFt	Maximum Width of Duct				
	0-12"	13-30"	31-54"	55-84"	185 and up
	26 ga	24 ga	22ga	20 ga	18 ga
Str Duct Only	54	80	84	98	112
10-20%	46	58	70	82	94
20-30%	41	57	62	73	83
30-40%	38	45	55	66	74
40-50%	35	43	50	58	66
Fittings only	24	30	35	41	47

3.   Rigid Board, 1" thick 1.50
     Rigid Board, 1-1/2 " thick 1.75
     Rigid Board, 2" thick 2.00
4.   1/2" thick blanket, 1-1/2 lb 85

Budget Figures
1.   Typical lining:
           1" thick, 1-1/2 lb, 30% markup, $1.75/sq
           1/2" thick, 1-1/2 lb, 30% markup, $1.46/sq

Chapter 15

Miscellaneous Labor Operations

DRAFTING AND SKETCHING LABOR

Miscellaneous labor covers additional operations which are not directly installation work. This may include shop drawings, sketching, cartage, testing and balancing, operation and maintenance manuals, sleeves and chases, excavating and backfilling, removal work, cutting and patching openings, etc.

Fully detailed, 1/4 inch scale, office prepared shop drawings with locations, elevations, outlet locations, fitting details, pipe and fitting lengths, walls, partitions, and reflected lights and beams all shown.

Labor

Labor includes the preparation of the shop drawings, revisions, field checks, making out shop fabrication tickets, and listing blankouts.

It is a function of the quantity of pieces of ductwork rather than the weight, fittings taking twice the total amount of time than straight pipe sections. It is also dependent on congestion in the ceiling space of all the mechanical, electrical work, on the complexity of the duct runs, and the extent of the architectural, structural complexity in the particular area.

Straight pipe ....10 hr/pc
Fittings ....20 hr/pc
Typical 50/50 mix, straight duct
    and fittings, by quantity pieces...........15 hr/pc

On a *pounds per hour* basis for an average mix of gauges:
    Mix 10-20% fittings by weight 23 5 lbs/hr
    Mix 20-30% fittings by weight 200 lbs/hr
    Mix 3040% fittings by weight 185 lbs/hr
    Mix 40-50% fittings by weight 175 lb/hr
    Pipe only 350 lbs/hr
    Fittings only 100 lb/hr

Correction Factors

Clear, open areas or straight runs	0.7
Congested ceiling spaces.................................	1.2
Equipment room..	1.2
Heavy duplications	0.8
Complete duplication of area	0.6

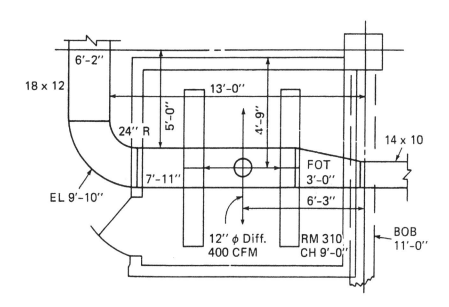

Budget Figures
　　Galvanized ductwork:
　　　　230 per pound
　　　　260 per square foot of ductwork

FIELD MEASURING AND SKETCHING LABOR

Final Duct Connections

　　Final duct and flexible connections to units, fans, louvers, etc.: 120lbs/hr or 50hr/pc, based on size of 48"x24".

Complete Duct Runs

　　Measure area, obstructions, sketch run, figure lengths, elevations, draw fittings—30% fittings by weight.

0-800 lbs	24 gauge avg 22 gauge avg	120 lbs/hr 200 lbs/hr	.25 hr/pc .25 hr/pc
800 lbs up	24 gauge avg 22 gauge avg	150 lbs/hr 300 lbs/hr	.2 hr/pc .2 hr/pc

Budget Figures for New Projects

　　Use 5% of total metal weight for finals and for measuring sheet metal specialties.

　　　　Ex. 50,000 lb job x 5% = 2500 lbs divided by
　　　　　　120 lbs/hr = 21 hrs total

　　Hence field measuring averages out to about 2000 lbs/hr or 2.50/lb based on total weight of job.

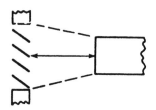

SERVICE

　　Two methods of calculating service and punch list work:
1.　2 hours per piece of equipment
2.　5% of the cost of heating, refrigeration, air handling equipment

ESTIMATING AIR TESTING AND BALANCING

Balancing Procedures

　　The labor figures in this section are based on the following air testing and balancing procedures.

1.　**Prepare test reports**
　　• Assemble plans, specs, equipment cuts, A_k factors.
　　• Survey and plan balancing.
　　• Determine instruments to use.
　　• Draw schematics.
　　• List outlets on outlet sheet.
　　• Determine A_k factors and calculate required velocities.

2.　**Start up**
　　• Check motor name plates, starter overload, heater size, fan, type wheel, rotation, drives, bearings, filters, etc.
　　• Check and set automatic dampers.
　　• Checkout dampers in duct runs, and at outlets and inlets. Check terminal units.
　　• Clean up debris.
　　• Turn fan on and take start up readings: amp, volts, fan rpm, total, CFM fan static pressure.

3.　**Balance duct runs and outlets** proportionately.

4.　**Reread equipment** and adjust as required.

5.　**Finalize reports**.

Instruments

　　The labor figures are also based on using the following instruments:

Volt-Ammeter	Pitot Tube
RPM Counter	Inclined Draft Gauges
Rotating Vane Anemometer	Magnehelic Gauges
Alnor Velometer	Thermometers

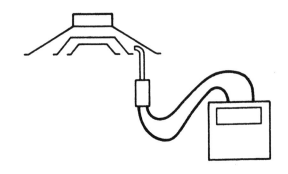

Start Up Labor

Includes checkout of equipment, setting automatic dampers, reading amps, volts, fan rpm, fan static pressure and total CFM.

Hours

Supply units	4.0
Multi-zone units (plus .5 hours per zone)	3.0
Centrifugal exhaust fans	3.0
Roof exhaust fans	2.0
Fan-coil units	2.0
Dust collectors	3.0

Correction Factors on start up labor   **Multipliers**

1. If no S.P. or total CFM readings
   are taken on supply units 0.70
2. Typical units 0.90

Balancing Outlets and Inlets

Based on 8 to 12 foot high ceilings, 2 to 3 passes, 3 to 4 readings each, using flow hood.

Diffusers	Smaller Simpler Systems	Larger Complicated Systems
6"-24"	0.35	0.50
26" up	0.50	0.70
Linear diffusers, per 5 foot length	0.40	0.60
Light troffers, per slot	0.30	0.40
Grilles, registers		
0-4 SF	0.35	0.50
4-8 SF	0.40	0.60
8 up SF	0.50	0.70
Exhaust hoods	0.40	0.60

Terminal Units

High pressure boxes:	access through ceiling	0.50
	access by crawling in ceiling	0.80
Induction units:	under windows	0.50
	in ceiling	1.00

Test Reports

Fill out equipment sheets, each system	0.20
Fill out and finalize outlet sheets, per outlet	0.12
Draw schematic layout, per outlet	0.05

Rule of thumb: all test reports, 10% of all field balancing work.

Correction Factors   **Multipliers**

1.	Ceiling height; 8 to 12 feet	1.00
	13 to 18 feet	1.15
	19 feet and up	1.25
2. Floor	1-5	1.00
	6-15	1.10
	16-30	1.20
	31 and up	1.25
3. Occupied areas		1.15
4. Hazardous industrial exhaust areas		1.25

Budget Figures

Includes all operations from preparing reports through startup, balancing and finalizing reports.

Figures are based on low pressure systems, single zone, low rise building, 8 to 12 foot high ceilings, 1 to 2 outlets per room, a typical mixture of outlet sizes and CFMs and that there are several systems on the job.

	PER OUTLET	
	Average Hours	Budget Each
Smaller, simpler systems:	0.75	$40
Larger, complicated systems:	1.00	$47

Example Testing and Balancing Estimate

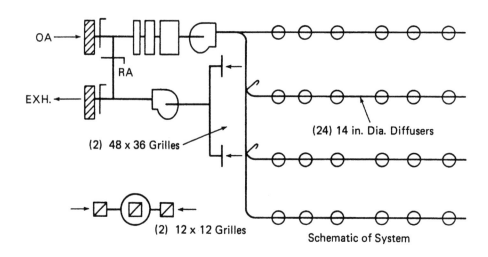

(24) 14 in. Dia. Diffusers

(2) 48 x 36 Grilles

(2) 12 x 12 Grilles

Schematic of System

Qty.	Item		Hours Each	Total
1	Supply unit	Start Up		4.0
1	Return air fan			3.0
1	Roof exhaust fan	↓		2.0
24	Diffusers (typical .9 x .5 = .45)	Bal.	.45	10.8
2	12x12 grilles		.5	1.0
2	48x36 grilles	↓	0.7	1.4
2	Pitot tube traverse in branch duct		.6	1.2
3	Equipment sheets	Test Reports	.2	.6
28	Outlet sheets		.12	3.4
28	Schematic drawing	↓	.05	1.4
1	Change RPM			.75
		Gross total		29.55
Small simple job .9 x 29.55		= Net total		26.6
Average per outlet 26.6 ÷ 28 =			.95hrs	

Estimating Ductwork Leak Testing
Medium or High Pressure Duct Runs

LABOR FOR TYPICAL DUCT RUN

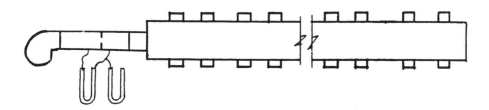

1.	Set up leak testing rig	1.0 hr
2.	Cap and seal ends of duct run, 1/2 hr each	1.0 hr
3.	Cap and seal branch collars, 16 at .2 hrs each	3.2 hr
4.	Check leakage with fan, walk run, seal	2.0 hr
5.	Retest	1.0 hr
	Total time for segment	8.2 hrs

ESTIMATING HYDRONIC BALANCING

Check Out Equipment Hours

Pumps: Check out pump itself, motor, starter, adjacent valves; read pressures, gpm, amps volts; adjust.	1.5 to 2.5
Chillers, Absorption Units	2.0 to 3.0
Cooling Tower	5.0 to 10.0
Central Cooling and Heating Coils	2.0 to 4.0

Balancing Terminals

Reheat Coils, Radiation Units	.75 to 1.25
Induction Units, Fan Coil Units	.75 to 1.25

Miscellaneous Labor Operations
Check Off List

DRAWING, CARTAGE
- ❒ Cartage
- ❒ Shop Drawings
- ❒ Field Sketching

TESTING, BALANCING, SERVICE
- ❒ Testing and Balancing
- ❒ Pressure Testing
- ❒ Monitoring
- ❒ Service
- ❒ Operation and Maintenance Manuals

MISCELLANEOUS
- ❒ Initial Set Up of Job
- ❒ Clean Up of Job

GENERAL CONSTRUCTION WORK
- ❒ Sleeves
- ❒ Chases
- ❒ Excavate
- ❒ Backfill

REMOVAL WORK
- ❒ HVAC Equipment
- ❒ Ductwork
- ❒ Piping
- ❒ Electrical
- ❒ Remove and Replace Partitions
- ❒ Ceilings
- ❒ Remove and Replace Doors
- ❒ Remove and Replace Windows
- ❒ Cut Openings
- ❒ Patch Openings
- ❒ Remove and Replace Ceilings
- ❒ Paint
- ❒ Protect Furnishings and Equipment
- ❒ Scrap items
- ❒ Clean Up

Section V

Piping Estimating

This page intentionally left blank

Chapter 16

Piping Estimating Basics

REQUIREMENTS OF A PROFICIENT ESTIMATOR

A proficient and reliable piping estimator must possess the following background knowledge, skills and abilities:

Estimating Principles and Procedures

He must follow sound efficient procedures for preparing estimates, such as:

- Become thoroughly familiar with the project, the types of systems and piping, valves etc., involved in the scope of work, etc. before starting a detailed takeoff.

- Be familiar with budget estimating piping systems, costs for different type buildings and systems based on: Cost per square foot of building or cost per ton of air conditioning, Amount of piping per square foot of building, The average size of piping on a job, Cost of piping per linear foot, Cost per square foot of building of piping.

- Know the major divisions of an estimate as described in "8 Facets of the Estimating Diamond" in Chapter 1:
    Equipment
    Piping, Valves
    Accessories, Specialties
    Special Labor
    Sub-Contractors End of Bid Factors (such as sales tax)
    Markups for Overhead and Profit

- Must be familiar with detailed scope of what is required in a piping estimate.

- Identify and highlight various types of pipe, fittings, valves, etc. on drawings before the takeoff.

- Follow systematic overall procedures as described in chapter 2.
    Study the plans and specs.
    Send out quotation requests
    Make Takeoffs and Extensions
    Summarize
    Recap and Markups

- Do constant systematic checking on each part as you go along and overall at the end. Double check everything.

Blue Print Readings

Must have the ability to read blue prints, recognize symbols, types of pipe lines, types of equipment and systems, etc.

Types of Piping Systems

Must be knowledgeable of the types of piping systems there are such as:
- Low, medium and high temperature **hot water** systems.
- Low, medium and high pressure steam systems.
- **Chilled water** cooling systems
- **Refrigeration** systems.
- **Hot** and **cold** water systems.
- **Oil** and **gas** piping.

He must not only recognize the various types of systems on plans, but he must know all of the components required in them, whether shown on plans or not.

They must know about different types of **piping system configurations** such as:
- 1,2,3 and 4 pipe systems
- Reverse and direct return
- Constant and variable volume
- Closed and open systems

Piping and Fitting Materials

A piping estimator must know about different types of pipe and fitting materials, manufacturing methods, types of fittings in each category and applications to systems.

Black Steel Pipe: Sch 20, 40 and 80, A53, A120, A106, Seam and Seamless Pipe, TC and PE ends
Black Fittings: Malleable, Butt Weld, Forged and Black Cast Iron Fittings
Copper Tubing: L, K, K ACR, Soft and Hard Tubing, plus DWV

Copper Fittings: Wrought and Cast Fittings
Pressure PVC: Sch 40, 80, Socket and Threaded Pipe
   and Fittings
Galvanized Pipe: Sch 40 and 80
Galvanized Fittings: Malleable
Cast Iron Soil Pipe: Hub and Spigot, No Hub

Material Applications

He must know the applications of different types of pipe and fitting materials to various systems:

Recirculating Water, 2500°F

For 2" diameter and under
   Black Steel A53 Seam, Sch 40 Threaded Pipe
   Malleable Threaded Fittings
For 2-1/2" to 12" diameter
   Black Steel A53 ERW Welded Pipe
   Butt Weld Fittings

Steam and Condensate

For 2" diameter and under, 90 lb
   Black Steel A53 Seam, Sch 40 Threaded Pipe
   Cast Iron Threaded Fittings
For 2-1/2" to 12", 250 lb
   Black Steel Standard Welded Pipe
   Wrought Steel Welded Fittings

Refrigerant

Copper L, Y, ACR Hard Tubing, Brazed
Wrought Copper Pressure Fittings, Brazed

Underground Water

Through 12" diameter, 350 lb
   Copper Y, Hard Tubing, Soldered
   Wrought Copper Pressure Fittings, Soldered

Portable Water Inside Building, 350 lb

Copper & Hard Tubing, Soldered
Wrought Copper Pressure Fittings, Soldered

Fittings

He must be familiar with different types of fittings and those available with all the different types of materials:

Long and Short Radius 90° and 45° **Elbows**
Straight and Reducing **Tees**
Concentric and Eccentric **Reducers**
Straight and Reducing 45° **Laterals**
Caps, Plugs, Unions, Adapters and Couplings
Weldolets, Threadolets and Sockolets
Threaded, Slip on and Welded Neck **Flanges**
Straight and reducing **Wye** Fittings, **DW**

Combination Fittings, **DW**
Bends, 1/8, 1/4, 1/6, etc., DAW
Straight, Reducing Taps, **Sanitary Crosses**, DWV
Traps, Cleanouts, DWV

Connections

He must have thorough knowledge of the different types of piping connections.

Steel:	Threaded
	Butt Weld
	Flanged
	Grooved
	Socket Weld
Copper:	95/5
	Brazed
PVC:	Solvent
	Heat Fusion
	Threaded
Cast Iron:	Soil Pipe
	Hub and Spigot
	No Hub

Hangers

He must be familiar with various types of hangers and supports such as:
   Clevis
   Roller
   Spring
   Riser
   Clamps

Labor

- He must know sources of labor such as MCA and NAPHCC association labor tables, other manuals available, cost records, etc.
- He must know the methods of estimating piping labor such as pipe per foot and fittings per piece or pure per joint labor method.
- He must apply labor multipliers when ever needed and do so with reasonable accuracy.

Piping Pricing

He must know sources of pricing such as piping supply houses, list pricing services such as Harrison, Allpriser, Trade Services, etc. He must be able to use quotations, pricing estimating manuals, etc.

Valves

He has to know about valves:

Standard Valves:
- Bronze, Iron, Steel, Plastic, etc.
- Gate, Ball, Globe, Butterfly, Check, Angle

Specialty Valves:
- Strainer
- Steam Traps
- Pressure Control
- Temperature Control
- Balancing
- Gas Valves
- Refrigeration
- Three or Four Way Combo check, gate and balancing

Specialties

He must know about specialties such as:
- Hot and Chilled Water
- Refrigeration
- Steam
- Plumbing
- Air Separators, Air Vents, Bleeders
- Roloirtrols (Combo air separator and strainer)
- Receivers, Sight Glasses, Dryers, Filters
- Vacuum Breakers, Drip Legs, Converters

Equipment

He must know about Equipment
- Pumps: Centrifugal, Inline, Single and Double Suction
- Boilers
- Unit Heaters
- Baseboard Heaters
- Chillers
- Compressors
- Condensers
- Cooling Towers
- Terminal Equipment
- Tanks

Gages

A piping estimator has to know about gages for temperature, pressure and flow readings.

Types of Insulation

He must be familiar with different types of insulation
- Fiberglass Per-form and Blanket
- Foam Plastic
- Calcium Silicate
- Eurethene Rigid
- Rubber Tubing, Foam

Wage Rates, Unions, Jurisdictions

He must know about wage rates, fringe benefits, federal, state and local taxes, insurance, etc.

He must be knowledgeable about union, trade and local labor jurisdictions and they must know about building codes.

Design

A piping estimator must have some familiarity with piping system design, such as typical flow rates, different systems, pressures and typical sizing for common parameters. He must be generally familiar with the selection of proper equipment.

Other Trades, Types of Buildings

They have to be familiar with other trades such as:
- Insulation
- Temperature Control
- Trenching
- Electrical

He must be familiar with all types of buildings, commercial, institutional, industrial, their general sizes, layout, etc. and with the sequence of general construction work.

Markups

A good piping estimator must be generally familiar with financial statements such as profit loss and balance sheets. He must be able to determine the correct markup for overhead and profit for their company and for the particular job he is bidding.

He should understand how overhead costs are pro-rated onto direct material and labor costs for different projects, for different levels of sales and overhead costs, for different ratios of material to labor, etc.

Skills, Traits Required

Estimating requires a host of skills, mathematical, mechanical, reading, writing, visualizing and drawing. It requires being methodical, analytical, strategical and realistic.

It absolutely demands that the estimator be reliable, that he be thorough in their understanding of the project, of its scope, in takeoffs, interpretations, extensions, summaries and recaps.

Thus, the knowledgeable, proficient and reliable piping estimator as described above will be able to produce complete and accurate estimates, which in turn becomes the required foundation blocks of successful contracting.

HOW TO ESTIMATE PIPING SYSTEMS

This section of piping basics will cover how to takeoff and extend the various piping components in a piping estimate, piping, fittings, accessories (flanges, hangers, connection materials, etc.) valves, specialties and equipment. It win cover what forms to use for larger projects and small jobs.

General Basis of Piping labor and Material Pricing

The installation labor and material pricing for piping, fittings, valves and specialties is a function of the following characteristics:

	Examples
Type Material:	(Black, copper, etc.)
Type Item:	(pipe, 90 Elbow, etc.)
Size:	(1/2", 3" diameter)
Type Connection:	(threaded, welded, flanged, etc.)
Type Hangers:	(band, ring, clevis, etc.)

Hence the labor and pricing tables, takeoff and extension sheet divisions are mostly divided up in the above manner. Hangers usually are not a fundamental category and standard lengths of pipe may be a factor with certain materials.

General Takeoff Approach and Job Breakdown

In making the takeoff, separate the job into different categories so that correction multipliers can be applied as required.

- Standard Floor Plans
- Congested Areas
- Crawl Space
- High Piping Runs
- Equipment Rooms
- Higher Floors

The takeoff should be separated into different materials and type systems.

e.g. Hot water heating, Black steel, Schedule 40, AW53
System P 1

Refrigeration System, Copper Tubing, Type L
System P2

Separate the takeoff into alternates so that alternate prices can be computed separately.

Equipment Hookups

Stop or start takeoffs at the equipment before taking off risers going up or down and hooking up to the piece of equipment. The hookups should be taken off separately and duplications should be treated as an assembly times the number pieces of equipment connected to. Then estimate the hookups separately.

Labor

The labor factors in the tables are based on normal working conditions.

- Non-productive supervision is not included.
- For Job Conditions correction multipliers should be applied.
- Normal Material Handling is included.

Taking off Piping

- Piping is taken off in footage and is listed separately.
- It is taken off per system, per material, type connection and diameter.
- Labor and pricing are computed on a per foot basis
- Piping comes in various lengths, e.g. Black A53 is 21 feet long in the smaller size range.
- Cuts are figured for pieces that are less than the standard length.
- Connection materials such as weld, solder, etc. or flanges, coupling are needed between pipe to pipe connections and must be counted and priced.
- Hanger sets are based on the footage of pipe and average spacing between sets, and must be counted and priced.
- The labor tables are generally based on an average of one piping connection for every five feet of pipe.
- Certain piping, such as cast iron hub and spigot may be labored and priced per standard length, for example, per two or five foot long standard section etc.

Taking off Fittings

- Fittings are taken on an "each" basis and listed separately from pipe on the takeoff sheets.
- They are taken off per system, per material, type connection and by diameter.
- Connections materials such as weld, solder, putty, etc. may be required at joints and need to be taken off, priced and labored.
- If the fitting is a flanged piping system or a connection to a piece of equipment with a flange, and the fittings are not flanged, then separate flanges that need to be taken off, price and labor them.

Large Projects; Takeoff Method and Extensions

On large projects where numerous takeoff sheets are required, separate piping takeoff and extension sheets should be used as opposed to a combination takeoff and extension sheet.

1. Piping takeoff sheets are used for taking off piping, fittings, valves, etc. Each column is headed with the particular diameter, except the left hand column which is used for listing different type of fittings, valves, etc.

2. After the takeoff is complete the footages of pipe for each diameter are totaled, as well as the total quantities of fittings, valves, etc. for each diameter being totaled.

3. The totals per type piece and diameter etc. are transferred to a piping extension sheet and summarized.

4. Unit labor and pricing are looked up in the tables in this manual and extended for a total labor and price for each type and size.

5. The columns are then totaled on the extension sheet and transferred to a piping recap sheet for the totals per type of material and connection.

6. The totals of each category are then transferred from the piping recap sheet to the estimate summary sheet for the job.

Small Job Takeoff and Extension Method

1. On small projects where maybe only one or a few takeoff and extension sheets are needed, the piping extension sheet is used for both the takeoff and extension of the piping, fittings, valves, etc.

2. In this case, sizes are simply listed in the left hand column, the type of item in the second column on the same line and the footages or quantities in the third column over.

3. After the takeoff is completed the footages and quantities are totaled in the middle column.

4. Unit material costs and erection labor hours are looked up, entered in the unit columns on the right hand side of the form and extended in the total columns.

5. The material and labor totals are then added up for either each category or for the entire sheet, and the numbers transferred to a job summary sheet accordingly.

SAMPLE ESTIMATE
Hot Water Heating System

Overview of Sample Piping Estimate

The sample piping system covers a hot water heating system for an office building with hot water heating coils in three air handling units, black steel threaded and welded piping, malleable threaded and butt welded fittings, bronze and iron body valves, control and balancing valves. It contains a hot water boiler, compression tank, pump, three way valve, rolairtrol air eliminator, thermometers and pressure gauges, plus other miscellaneous items.

Purpose of Forms

Purpose of Forms

Forms are an indispensable aid and guide to organized, efficient and thorough estimating. They help control the proper sequence of estimating work, continually remind you of what information is needed, lead you logically through calculations and as a result your bids will be more complete and correct.

Job Description and Budget Costs Form (See form samples in chapter 3)

1.   Budget estimate prices to determine if it should be bid or not, and as a check price against the detailed estimate after the bid is complete.

2.   Approximate heating and cooling loads and rough out total piping linear footage for check of detailed piping takeoff.

3.   Record the key characteristics of the type of system involved.

Piping Takeoff Sheet

The piping takeoff sheet can be used for piping, fittings, valves, gages, strainers, etc on larger type projects. Estimating piping labor by the hours per foot, and estimating fitting, valve, gage, strainer, etc. labor by the piece, is the most accurate and clearest method available for contractors. The takeoff involves listing the diameter, type, quantities on fittings, valves, etc. and lengths on pipe.

Piping Extension Sheet

The extension of material involves totaling piping footages per line, entering the labor per foot and extending totals. The extension of fitting, valve, etc. labor involves totaling the quantity of pieces, looking up and entering labor hours per piece and multiplying out for the totals per line. The same procedure is followed for pricing. After the lines are extended the columns are totaled.

Combined Piping Takeoff and Extension Sheet

The combined piping and takeoff sheet is more practical for smaller projects involving fewer items and one or a few sheets. Sizes are listed horizontally in the left hand column, rather than in the column headings.

Equipment Takeoff and Extension Sheet

List all equipment, quantities, sizes, capacity, description on this combined sheet and enter purchase costs from quotations, pricing sheets if available. Enter labor to install equipment using labor tables in this manual or other sources.

Quantity Takeoff and Extension Sheet (See form samples in chapter 3)

The quantity takeoff sheet is a general form for taking off and listing types, sizes, quantities, etc. of the various items required in a bid other than ductwork or piping, for extending the material amounts, labor, costs, etc. and summations.

Estimate Summary and Extension Sheet (See form samples in chapter 3)

The summary sheet is used as a line item summary of all the major grouping of different items included in the estimate, from ductwork and other takeoff sheets, etc. It should be divided into the major divisions of a bid, quoted equipment, ductwork, piping, specialties and accessories, miscellaneous labor, etc.

The total amounts of material quantities, labor, etc. are transferred from duct and piping takeoff sheets, piping extension sheets, quantity takeoff sheets etc. to this summary sheet.

Bid Recap Sheet (See form samples in chapter 3)

1.   Recap the job totals of direct costs on labor and prices on raw materials, equipment and sub contractors, and to total them.
2.   Put markups on each group and total the overhead markup.
3.   Put a profit markup on the labor, raw material, equipment and sub-contractor groups.
4.   Total everything for a bottom line bidding price.

Calculating Labor Costs Per Hour Form (See form samples in chapter 3)

This form insures that all the components of the wage rate which include, base wage rate, normal union fringe benefits, federal and state payroll taxes, insurance's and dues, are covered in the rate used in a bid.

Telephone Quotations Form (See form samples in chapter 3)

The telephone quotation form is for recording quotations which come over the phone, in an organized, complete and readable fashion. It includes a check off list on the bottom of critical aspects of a quote such as, if they meet plans and specification requirements, addendum's, taxes, freight, lead times, etc. A box is provided for exceptions on what is not included.

Bidding Record Form (See form samples in chapter 3)

The purpose of the bidding record form is to have a written record of to whom the phone bids were given, what the amount was, what the inclusions and exclusions were and what the plans, specifications and addenda of the bids were based on.

Specifications on Sample Job

IBM Sales Office

A.   Related Documents

The general provisions of the contract, the general conditions and supplementary conditions of these specifications plus the A/A document A201-1976 "general conditions" apply to the work in this specification.

B.   Scope of Work

Scope of work to include, but not be limited to the following:

Equipment
1. Boiler
2. Roloirtrol
3. Centrifugal Pump
4. Hot Water Coils
5. Chemical Pot Feeder
6. Compression Tank
7. Control Dampers

Piping
8. Black Steel, Sch 40, A53, Threaded
9. Black Steel, Sch 40, A53, Welded

Fittings
10. Black Malleable Threaded
11. Black Steel Butt Welded

Valves
12. Bronze, Threaded
13. Iron Body

Specialties Valves
14. Control Balancing

Specialties
15. Strainers
16. Drains

Gauges
17. Thermometers
18. Pressure

Miscellaneous
19. Test piping systems

Sub Contractors
20. Temperature Controls
21. Insulation

C.   Work Not Included
1. Painting
2. Power wiring to mechanical equip.
3. Structural steel openings
4. Concrete Pads
5. Starters

Sample Hot Water Heating System

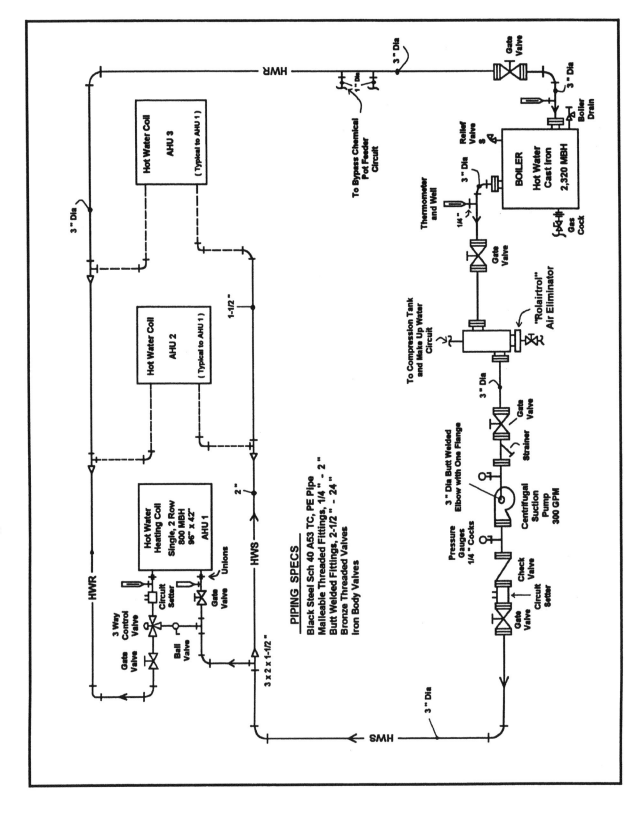

Piping Takeoff Sheet

Job *Heating System* Drawing *M-1* System *P1* Factors_____ Date_____
Page ___*1*___

Type ___*Black, 40, A53, PE & TC*___ Connection *Thrd 1", Weld 2.5"* ☐ Insulation

Valves _____

Item	Diameter			
	1"	1 1/2"	2"	3"
Pipe	20-22-3-3-3-3	22-13-14-2-3	14-7-3-3-21	2-3-8-7-3-3
	2-3-2-3-4-3-3	8-10-2-2-2	5-3-9-3-3	3-3-3-48-3
	3-2-2-2 (80)	(78)	(71)	(83)
Nipples				
12" Lengths	~~卌 卌 卌 卌~~			
	~~卌~~ ////			
Fittings				
90 Elbows	~~卌 卌 卌~~	~~卌~~ ////	~~卌 卌~~ //	////
Straight Tees	1x1x1 - (///)	1.5x1.5x1.5 - (//)	2x2x2 - (/)	3x3x3 - (/)
Reducing Tees	1.5x1x1 - (/)			3.5x1.5x1.5 - (/)
Unions	~~卌~~	/	//	
Threadolets	//			/
Flanges				~~卌 卌~~
Valves				
Gate	/			/
Ball	~~卌~~ /	/	/	
Butterfly				//
Globe	/			
Triple Duty				/
Thermostatic Radiator	/			
2 Way Control	/			
3 Way Control			/	
Circuit Setter	//			
Draw Valves	///			
Pressure Relief Valves	/			
Gages				
Thermometer & Well	////			
Pressure Gage	//			
1/4" Cock	//			
Airtrol Tank	/			
Sight Glass	/			

Sample piping takeoff sheet for larger jobs requiring numerous takeoff sheets. Piping taken off in footage and continuous lengths between connections. Fittings, valves and gauges taken off in quantity of pieces.

Piping Extension Sheet

Date_____

Job *Heating System* Drawing *M-1* System *P1* Factors_____ Page ____1____

Type ___*Black, 40, A53, PE & TC*___ Connection *Thrd 1", Weld 2.5"*____ ☐ Insulation

Valves _____

Diameter	Item	Quantity or Linear Feet	Total	Material Cost		Erection Labor		
				Unit	Total	Eqv Joints	Unit Hr	Total Hr
1"	Pipe		80'	$1.83	$146.40		0.07	5.60
1-1/2"			78'	2.83	220.74		0.08	6.24
2"			71'	3.53	271.93		0.09	6.39
3"			83'	6.97	578.51		0.24	19.92
1"	Nipples 12"		29	4.10	115.90		0.07	2.03
1"	90 Elbows		15	2.08	31.20		0.64	9.60
1-1/2"			9	4.47	40.23		0.80	7.20
2"			12	6.57	78.84		0.88	10.56
3"			4	24.89	99.56		2.40	9.60
1"	Straight Tees		3	3.22	9.66		0.96	2.88
1-1/2"			2	6.50	13.00		1.20	2.40
2"			1	9.44	9.40		1.32	1.32
3"			1	32.28	32.28		3.60	3.60
1.5"x1"x1"	Reducing Tees		1	7.01	7.01		1.20	1.20
3"x1.5"x1.5"			1	35.33	35.33		3.60	3.60
1"	Unions		5	5.99	29.95		0.64	3.20
1-1/2"			1	10.07	10.07		0.80	0.80
2"			2	12.15	25.50		0.88	1.76
1"	Threadolets		3	7.25	21.75		0.64	1.92
3"	Flanges		10	13.80	135.00		1.20	12.00
		Totals	99		$1,918.30			111.82
		Valves - Pg. 2			1769.69			24.28
		Total			$3,687.99			136.10
		30% Supplier Discount *(ex. only)			x .70			
		Net Material Cost			$2,581.59			

Sample piping extension sheet when used in conjunction with separate piping takeoff sheets on larger jobs.

Note: The unit prices above may be slightly different than the ones in this manual due to continuous updating.

* The supplier discount on the total material at the bottom of the material cost column must be determined by the contractor.

For a 30 percent discount from list price, a multiplier of 70 percent is used:
    (1.0 - .30 discount = .70 multiplier)   [example of calculation only]

Piping Extension Sheet

Date _____

Job *Heating System* Drawing *M-1* System *P1* Factors _____ Page _____2_____

Type *Bronze 2", Iron 2.5"* Connection *Thrd 2", Flangd 2.5"* ☐ Insulation

Diameter	Item	Quantity or Linear Feet	Total	Material Cost		Erection Labor		
				Unit	Total	Eqv Joints	Unit Hr	Total Hr
1"	Gate Valve		1	$24.15	$24.15		0.64	0.64
3"			1	127.05	127.05		1.46	1.46
1"	Ball Valve		6	10.88	65.28		0.64	3.84
1-1/2"			1	18.98	18.98		0.80	0.80
2"			1	30.00	30.00		0.88	0.88
3"	Butterfly		2	35.09	70.18		1.46	2.92
1"	Globe		1	45.00	45.00		0.64	0.64
3"	Triple Duty		1	293.40	293.40		1.46	1.46
1"	Therm. Rad.		1	147.50	147.50		0.64	0.64
1"	2-Way Contrl		1	113.00	113.00		0.64	0.64
2"	3-Way Contrl		1	259.00	259.00		0.88	0.88
1"	Circuit Setter		2	40.00	80.00		0.64	1.28
1"	Drain Valve		3	24.15	72.45		0.64	1.92
1"	P.R.V.		1	81.00	81.00		0.64	0.64
1"	Therm. & Well		4	50.00	200.00		0.64	2.56
1"	Press. Gage		2	16.50	33.00		0.64	1.28
1/4"	Cock		2	6.25	12.50		0.26	0.52
1"	Airtrl Tnk Ft.		1				0.64	0.64
1"	Sight Glass		1	97.20	97.20		0.64	0.64
		Total	33		$1,769.69			24.28
		30% Supplier Discount *(ex. only)			x .70			
		Net Material Cost			$1,238.78			

Note: The unit prices above may be slightly different than the ones in this manual due to continuous updating.

* The supplier discount on the total material at the bottom of the material cost column must be determined by the contractor.

For a 30 percent discount from list price, a multiplier of 70 percent is used:
(1.0 - .30 discount = .70 multplier)   [example of calculation only]

Alternate Method
Combination Piping Takeoff and Extension for Small Jobs

Job __Atrium Restaurant__ Drawing __M-1__ System __Gas__ Factors_____ Date_____

Page_____

Type Pipe __Black Iron_____ Connection __Screwed_____ ☐ Insulation

Diameter	Item	Quantity or Linear Feet	Total	Material Cost		Erection Labor		
				Unit	Total	Eqv Joints	Unit Hr	Total Hr
3/4"	Pipe	2-6-2-9-6-2-4						
		6-2-8-2-8-2-8						
		2-8	79'	0.86	68.00	17	0.28	4.76
1"		30	30	1.18	35.40	3	0.34	1.02
1-1/2"		23	23	1.78	41.00	2	0.44	0.88
2"		2-10	12	2.49	30.00	2	0.64	1.28
		(total pipe)	144'					
2"	90 Elbow	// (pieces)	2	3.50	7.00	4	0.64	7.04
3/4"		𝟋𝑯𝑳 𝟋𝑯𝑳 ////	14	0.60	16.80	28	0.28	7.84
2x2x1	Tee	/	1	-	5.05	3	0.64	1.92
1.5x1.5x3/4"		/	1	-	4.63	3	0.44	1.32
3/4"	Union	𝟋𝑯𝑳 ///	8	0.96	7.68	16	0.28	4.48
2"		/	1	-	5.90	2	0.64	1.28
1"	Couplings	/	1	-	1.35		-	-
1-1/2"		/	1	-	2.20		-	-
		(total fittings)	29					
3/4"	Gas Stops	𝟋𝑯𝑳 //	7	4.60	32.20	2	0.28	0.56
1"	Hangers		15	1.26	18.90		0.50	7.50
		Totals			$276.11			40.00
		30% Supplier Discount			x.70			
					$193.28			

Sample piping extension sheet being used for both the takeoff and extension on small jobs. The unit prices above may be slightly different than the ones in this manual due to continuous updating.

Equipment Takeoff and Extension Sheet

Date_____

Job *Heating System* Drawing **M-1** System __P1__ Factors_____ Page____1____

Quantity	Item	Price		Labor Man Hours	
		Unit	Total	Unit Hr	Total Hr
3	HW Heating Coil	$520	$1,560	20.0	60.0
	96" x 42", 800 MBH				
	Single, 2 Row				
1	Hot Water Boiler, Cast Iron		$1,801		74.0
	2,320 MBH, Gas Fired				
1	Centrifugal Suction Pump		$2,845		10.0
	300 GPM, 3500 RPM				
1	Bypass Chemical Pot Feeder		**		1.6
	1" Diameter				
1	Compression Tank		**		3.0
	Total		$		148.6
	** Call for Quote				

Note: The unit prices above may be slightly different than the ones in this manual due to continuous updating.

* The supplier discount on the total material at the bottom of the material cost column must be determined by the contractor.

For a 30 percent discount from list price, a multiplier of 70 percent is used:
(1.0 - .30 discount = .70 multplier)   [example of calculation only]

This page intentionally left blank

Chapter 17

Pressure Pipe, Fittings and Insulation

This chapter covers pressure pipe and fittings used for the following type systems:

- Hot Water
- Chilled Water
- Refrigeration
- Steam and Condensate
- Oil and Gas
- Hot and Cold Potable Water

This chapter further covers all types of the following pressure pipe and fitting materials:

- Black Steel
- Galvanized
- Carbon Steel
- Malleable
- Copper
- Pvc

These tables in this chapter also covers all the normal type **fittings** involved in pressure type systems such as radius elbows, street elbows, tees, laterals, crosses, caps, plugs, unions, couplings, weldolets, etc.

Standard type **connections** are covered, such as threaded, welded, grooved, soldered, solvent, etc.

Flanges such as threaded, slip on and welded neck are also covered.

Pricing for different types of **pipe hangers** such as band, clevis, roller, and riser clamps are included, plus standard hanger spacing, rod diameter tables, concrete inserts and anchors.

Piping sleeves and cutting round openings are covered.

At the end of this chapter are tables on **insulation** as follows:

Fiberglass Blanket
Fiberglass Preformed
Foam Plastic
Calcium Silicate
Rigid Eurethene
Foam Rubber Tubing

Units

All **diameters** are in inches.

Labor

The labor productivity rates in the tables are what the average mechanic under average conditions can erect on lower floors of construction projects. However, they may require the application of **labor correction** factors for variable conditions such as upper floors, higher piping runs, congested spaces, etc.

The labor in the tables are in hours per piece and hours per foot for pipe.

Pricing and Discounts

The pricing in the tables are based on the typical manufacturers list prices and the contractor's supplier discounts must be applied.

Appropriate price discounts can range from roughly 10 percent to 80 percent dependent on the following:

- Supplier discount structure
- Who manufacturer is
- Size of batch involved
- Availability
- Shipping distance, costs

Not All List Prices Are Publicly Available

Where there are no prices listed in tables, they are not usually publicly available from the manufacturer. If there is neither labor nor prices listed for a particular diameter in the tables, the particular item may not be available or it may not be a commonly used item.

Caution With List Prices

Pricing can sometimes vary significantly because of changes in commodity material pricing such as copper, etc. Periodic checks of current list prices should be made.

If in doubt about your discounts from list prices, *check with your supplier* as to what they are for the various categories of items—or request a *quotation*.

Automatic Updating
Available in Wendes Computer System

Automatic updating of list pricing is available in the *Wendes Computerized Piping Estimating System.* Call (847) 808-8371 for more information.

Terminology Abbreviations

TC pipe: Cut and threaded.
PE pipe: Plain end.
Thrd: Threaded connections.
Weld: Welded connections.
Grvd: Grooved connections.

Black BW A53 Schedule 40 Pipe
Labor

	Hours per Foot		
	TC Pipe	PE Pipe	
Dia	Thrd	Weld	Grvd
1/8	0.04	0.05	—
1/4	0.05	0.06	—
3/8	0.05	0.06	—
1/2	0.06	0.06	—
3/4	0.06	0.06	0.06
1	0.06	0.08	0.06
1-1/4	0.08	0.08	0.07
1-1/2	0.08	0.10	0.08
2	0.10	0.13	0.09
2-1/2	0.13	0.15	0.10
3	0.16	0.17	0.12
3-1/2	0.18	0.19	0.14
4	0.20	0.23	0.16
5	—	0.28	0.20
6	—	0.32	0.25
8	—	0.38	0.30
10	—	0.45	0.35
12	—	0.56	0.40
14	—	—	0.42
16	—	—	0.45
18	—	—	0.48
20	—	—	0.50
24	—	—	0.57

Labor Factor

Black BW A53 Schedule 80 Pipe	
Threaded TC Pipe over 2" Dia.	1.15
Welded PE Pipe all Diameters	1.20

Black Malleable Standard Fittings
Threaded Connection Labor

Dia	Threaded Labor Hours Each							
	90 Elbow	45 Elbow	Tee	Wye	Cross	Cap	Coupling	Brass Union
1/8	0.14	0.09	0.16		0.25	0.13	0.15	0.23
1/4	0.21	0.10	0.25	—	0.31	0.14	0.16	0.27
3/8	0.28	0.22	0.38	0.45	0.45	0.15	0.21	0.31
1/2	0.35	0.35	0.51	0.51	0.66	0.20	0.35	0.35
3/4	0.46	0.46	0.70	0.70	0.95	0.25	0.46	0.50
1	0.60	0.60	0.87	0.87	1.13	0.30	0.60	0.56
1-1/4	0.70	0.70	1.00	1.00	1.32	0.36	0.70	0.63
1-1/2	0.75	0.75	1.10	1.10	1.45	0.40	0.75	0.70
2	0.90	0.90	1.35	1.35	1.75	0.46	0.90	0.95
2-1/2	1.47	1.47	2.18	2.18	2.88	0.75	1.46	1.47
3	1.76	1.76	2.62	2.64	3.46	0.90	1.74	1.76
3-1/2	2.06	2.06	3.06	—	—	1.04	—	—
4	2.37	2.37	3.53	3.55	4.64	1.19	2.32	—
5	3.84	3.84	5.70	—	—	1.93	—	—
6	4.65	4.65	6.89	—	—	2.32	—	—

	Labor Factor
Black malleable standard fitting threaded XH	1.03

Black Malleable Standard Fittings
Grooved Connection Labor

22-1/2 Dia	Grooved Labor Hours Each									
	90 Elbow	45 Elbow	Lateral Elbow	Tee	Female Wye	Cross	Std Flex Adapter	Cap	#150 Coupling	Flange
3/4	0.35	0.35	0.42	0.50	0.88	0.98	0.18	0.18	0.03	—
10.42	0.42	0.45	0.60	0.88	0.98	0.22	0.22	0.04	—	
1-1/4	0.45	0.45	0.50	0.68	0.88	0.99	0.25	0.24	0.05	—
1-1/2	0.50	0.50	0.60	0.70	0.89	1.00	0.27	0.25	0.05	—
20.60	0.60	0.58	0.90	0.89	1.15	0.32	0.30	0.06	0.05	—
2-1/2	0.65	0.65	0.70	0.90	0.87	1.16	0.30	0.31	0.08	0.07
3	0.70	0.70	0.80	1.05	1.05	1.40	0.35	0.35	0.09	0.08
4	0.95	0.95	0.93	1.45	1.42	1.85	0.50	0.47	0.10	0.10
5	1.45	1.40	1.40	2.15	2.12	2.75	—	0.70	0.12	0.12
6	1.75	1.70	1.70	2.62	2.60	3.35	—	0.85	0.20	0.15
8	2.45	2.30	2.28	3.60	3.55	4.45	—	1.18	0.25	0.20
10	3.30	3.00	2.90	4.85	4.55	5.75	—	1.53	0.35	0.20
12	4.00	3.30	3.00	5.30	4.75	5.95	—	1.67	0.45	0.50
14	4.35	3.73	—	5.70	5.45	6.35	—	—	0.70	0.70
16	5.05	4.10	—	6.14	5.80	6.55	—	—	0.82	0.85
18	5.30	4.55	—	7.38	7.02	8.65	—	—	0.84	0.90
20	6.15	5.30	—	8.25	8.05	9.95	—	—	1.6	1.12
24	6.95	6.80	—	9.90	9.35	10.93	—	—	1.62	1.53

Carbon Steel Butt Weld Standard Fittings
Welded Connection Labor

Dia	Welded Labor Hours Each				
	90 LR Elbow	45 LR Elbow	Tee	Cross	Cap
1/2	0.57	0.57	0.84		0.31
3/4	0.84	0.84	1.24		0.44
1	1.11	1.11	1.65	—	0.57
1-1/4	1.39	1.38	2.06	2.73	0.71
1-1/2	1.65	1.64	2.47	3.27	0.84
2	2.20	2.18	3.29	4.35	1.11
2-1/2	2.56	2.55	3.85	5.15	1.25
3	2.70	2.65	4.05	5.35	1.35
3-1/2	3.10	3.05	4.65	6.15	1.54
4	3.35	3.32	5.03	6.70	1.65
5	3.92	3.85	5.85	7.75	1.93
6	4.40	4.25	6.55	8.60	2.15
8	5.65	5.40	8.40	10.90	2.70
10	7.30	6.90	10.65	13.90	3.45
12	9.05	8.40	13.15	16.92	4.20
14	10.70	9.85	15.90	19.80	4.90
16	12.50	11.40	18.00	22.90	5.65
18	14.34	13.00	20.40	25.95	6.40
20	15.85	14.60	23.15	28.85	7.20
22	17.50	16.25	25.35	—	8.00
24	19.15	17.95	25.25	34.65	8.85

	Labor Factor
Carbon steel buttweld XH	1.12

Carbon Steel Welding Flanges
Threaded and Welded Connection Labor
to Attach to Pipe or Fitting

Dia	Threaded	Flanges Raised Face #150			
		Weldneck	Slip-on	Socket	Blind
1/2	0.52	0.67	0.67	0.58	0.27
3/4	0.55	0.70	0.70	0.60	0.29
1	0.60	0.85	0.85	0.68	0.30
1-1/4	0.65	1.00	1.00	0.78	0.32
1-1/2	0.68	1.12	1.12	—	0.33
2	0.80	1.45	1.45	1.10	0.35
2-1/2	1.00	1.65	1.65	1.20	0.40
3	1.15	1.75	1.75	1.32	0.42
3-1/2	1.40	2.15	2.15	1.65	0.70
4	1.70	2.50	2.50	1.95	0.75
5	2.30	2.80	2.75	2.20	0.90
6	2.65	3.00	2.90	2.27	0.95
8	3.45	3.70	3.60	2.75	1.10
10	4.55	4.85	4.75	3.70	1.70
12	5.35	5.75	5.60	4.30	2.35
14	6.25	6.80	6.62	—	2.85
16	7.30	7.90	7.55	—	3.55
18	8.50	8.85	8.62	—	4.05
20	10.00	10.40	10.25	—	5.00
24	11.70	13.00	12.55	—	5.80

Galvanized BW A53 Schedule 40 Pipe

Labor

	Hours per Foot		
	TC Pipe	PE Pipe	
Dia	Thrd	Weld	Grvd
1/8	0.04	—	—
1/4	0.05	—	—
3/8	0.05	—	—
1/2	0.06	—	—
3/4	0.06	—	0.06
1	0.06	—	0.06
1-1/4	0.08	—	0.07
1-1/2	0.08	—	0.08
2	0.10	—	0.09
2-1/2	0.13	0.15	0.10
3	0.16	0.17	0.12
3-1/2	0.18	0.19	0.14
4	0.20	0.23	0.16
5	—	0.28	0.20
6	—	0.32	0.25
8	—	0.38	0.30
10	—	0.45	0.35
12	—	0.56	0.40
14	—	—	0.42
16	—	—	0.45
18	—	—	0.48
20	—	—	0.50
24	—	—	0.57

Galvanized BW A53 Schedule 80 Pipe:		Labor Factor
Threaded TC Pipe	0 - 2″	same
	2-1/2 - 4″	1.15
Welded PE Pipe		1.20

Galvanized Malleable Standard Fittings
Threaded Connection Labor

| Dia | Threaded Labor Hours Each | | | | | | | |
	90 Elbow	45 Elbow	Tee	Y Branch	Cross	Cap	Coupling	Brass Union
1/8	0.14	0.09	0.16	—	0.25	0.13	0.15	0.23
1/4	0.21	0.10	0.25	—	0.31	0.14	0.16	0.25
3/8	0.28	0.22	0.38	0.45	0.45	0.15	0.21	0.30
1/2	0.35	0.35	0.51	0.51	0.66	0.20	0.35	0.35
3/4	0.46	0.46	0.70	0.70	0.95	0.25	0.46	0.50
1	0.60	0.60	0.87	0.87	1.13	0.30	0.60	0.54
1-1/4	0.70	0.70	1.00	1.00	1.32	0.36	0.70	0.60
1-1/2	0.75	0.75	1.10	1.10	1.45	0.40	0.75	0.70
2	0.90	0.90	1.35	1.35	1.75	0.46	0.90	0.95
2-1/2	1.45	1.45	2.15	2.15	2.85	0.73	1.45	1.45
3	1.75	1.75	2.60	2.62	3.45	0.90	1.72	1.74
4	2.35	2.30	3.50	3.52	4.62	1.17	2.30	—
5	3.85	3.75	5.70	—	7.53	1.90	—	—
6	4.6	4.58	6.86	—	9.10	2.30	—	—

	Labor Factor
Galvanized malleable standard fitting threaded XFL 2-1/2″ - 6″	1.03

Copper L, K, M and ACR L-Hard Tubing
Soldered Connection Labor

Dia	K		L		M		ACR L-Hard	
	Hard	Soft	Hard	Soft	Hard	Soft	Dia	Solder
1/4	0.03	0.03	0.03		—	—	—	—
3/8	0.04	0.04	0.03	—	.0.03	0.03	3/8	0.02
1/2	0.04	0.04	0.03	0.03	0.03	0.03	1/2	0.03
5/8	—	0.04	—	—	0.03	0.03	5/8	0.04
3/4	0.05	0.05	0.04	0.04	0.03	0.03	3/4	0.05
1	0.05	0.05	0.05	0.05	0.05	0.05	7/8	0.05
1-1/4	0.05	0.06	0.05	0.05	0.05	0.05	1-1/8	0.05
1-1/2	0.06	0.06	0.06	0.06	0.05	0.05	1-3/8	0.06
2	0.07	0.07	0.06	0.06	0.06	0.06	1-5/8	0.07
2-1/2	0.08	0.10	0.07	—	0.07	0.07	2-1/8	0.07
3	0.10	0.13	0.10	—	0.10	0.10	2-5/8	0.08
3-1/2	0.11	0.15	—	—	—	—	3-1/8	0.10
4	0.13	0.16	0.12	—	0.11	0.11	4-1/8	0.15
5	0.15	—	0.15	—	0.15	0.15	—	—
6	0.20	—	0.18	—	0.17	0.17	—	—
8	0.30	—	0.25	—	0.23	0.23	—	—

Labor Hours Per Foot

Wrot Copper Pressure Standard Fittings
Soldered Connection Labor

Dia	90 Elbow	45 Elbow	Tee	Adapter	Cap	Coupling	Union	#125 Flange
1/8	0.13	0.13	0.20	0.13	0.05	0.13	0.13	—
1/4	0.16	0.16	0.26	0.14	0.08	0.16	0.16	—
3/8	0.22	0.22	0.35	0.15	0.10	0.22	0.22	—
1/2	0.27	0.27	0.40	0.15	0.15	0.27	0.27	0.43
5/8	0.30	0.30	0.46	—	0.17	0.30	0.30	—
3/4	0.35	0.35	0.53	0.17	0.18	0.35	0.35	0.44
1	0.41	0.41	0.63	0.20	0.21	0.41	0.41	0.47.
1-1/4	0.50	0.50	0.70	0.22	0.23	0.50	0.50	0.50
1-1/2	0.51	0.51	0.75	0.25	0.25	0.51	0.51	0.52
2	0.58	0.58	0.85	0.30	0.30	0.58	0.58	0.60
2-1/2	0.94	0.94	1.40	0.70	0.45	0.94	0.94	0.82
3	1.15	1.15	1.70	0.86	0.56	1.15	1.15	0.85
3-1/2	1.30	1.30	1.95	—	0.67	1.30	—	—
4	1.50	1.50	2.25	1.15	0.75	1.50	—	1.45
5	2.82	2.82	4.25	—	—	2.82	—	2.25
6	3.40	3.40	5.12			3.40	—	2.50
8	4.62	4.62	6.93			4.62	—	3.18

PVC Pressure Pipe, Schedule 40
Threaded and Solvent Connection Labor

Dia	Labor Hours per Foot	
	Thrd	Solvent
1/8	0.02	0.02
1/4	0.02	0.02
3/8	0.02	0.02
1/2	0.02	0.02
3/4	0.03	0.03
1	0.03	0.03
1-1/4	0.04	0.04
1-1/2	0.04	0.04
2	0.04	0.04
2-1/2	0.05	0.05
3	0.06	0.05
3-1/2	0.07	0.05
4	0.08	0.06
5	0.09	0.06
6	0.12	0.07
8	0.15	0.09
10	0.20	0.11
12	0.25	0.14
14	0.28	0.22
16	0.32	0.25

PVC Pressure, Schedule 40 Standard Fittings
Solvent Connection Labor

	Solvent Labor Hours Each								Slip
	90 D	45 D	30 D	22-1/2 D	15 D	11-1/4 D			
Dia	Elbow	Elbow	Elbow	Elbow	Elbow	Elbow	Tee	Wye	Cross
1/4	—	—	—	—	—	—	—	—	—
3/8	0.10	—	—	—	—	—	0.14	—	—
1/2	0.10	0.10	—	—	—	—	0.14	—	0.17
3/4	0.11	0.11	—	—	—	—	0.15	—	0.18
1	0.12	0.12	—	—	—	—	0.15	—	0.19
1-1/4	0.13	0.13	—	—	—	—	0.18	—	0.23
1-1/2	0.14	0.14	—	—	—	—	0.20	—	0.27
2	0.17	0.17	0.17	0.17	0.17	0.17	0.24	—	0.32
2-1/2	0.18	0.18	—	—	—	—	0.27	—	0.35
3	0.22	0.22	0.22	0.22	0.22	0.22	0.30	0.30	0.42
4	0.30	0.28	0.28	0.28	0.28	0.28	0.40	0.40	0.55
5	0.41	0.39	0.39	0.39	0.39	0.39	0.58	0.58	—
6	0.52	0.50	0.50	0.50	0.50	0.50	0.75	0.75	1.14
8	0.80	0.76	0.76	0.76	0.76	0.76	1.18	1.18	1.80
10	1.23	1.13	1.13	1.13	1.13	1.13	1.78	1.78	2.40
12	1.50	1.38	1.38	1.38	1.38	1.38	2.70	2.70	2.80
14	1.91	1.75	1.75	1.75	1.75	1.75	3.40	3.40	3.50
16	2.40	2.18	2.18	2.18	2.18	2.18	4.24	4.24	4.40

PVC Pressure, Schedule 40 Standard Fittings
Solvent Connection Labor

Dia	Adapter SxF	Adapter SxM	Slip Cap	Female Thrd Cap	Male Thrd Plug	Fitting Plug	Coupling SxS	Union SxS	Slip-on Flange
				Solvent Labor Hours Each					
3/8	0.16	0.16	0.06	0.06	0.02	0.02	0.10	—	—
1/2	0.18	0.18	0.06	0.06	0.02	0.02	0.10	0.10	0.25
3/4	0.22	0.22	0.07	0.07	0.02	0.02	0.11	0.11	0.25
1	0.27	0.27	0.07	0.07	0.02	0.02	0.11	0.11	0.27
1-1/4	0.30	0.30	0.08	0.08	0.02	0.02	0.12	0.12	0.27
1-1/2	0.33	0.33	0.09	0.09	0.02	0.02	0.14	0.14	0.27
2	0.40	0.40	0.10	0.10	0.02	0.02	0.16	0.16	0.28
2-1/2	0.62	0.62	0.11	0.11	0.03	0.03	0.18	—	0.30
3	0.70	0.70	0.12	0.12	0.03	0.03	0.20	—	0.30
4	1.00	1.00	0.15	0.15	0.03	0.03	0.28	—	0.60
5	1.28	1.28	0.20	0.20	—	—	0.40	—	—
6	1.55	1.55	0.26	0.26	—	—	0.48	—	0.65
8	2.52	2.52	0.40	—	—	—	0.75	—	0.74
10	—	—	—	—	—	—	0.95	—	—
12	—	—	—	—	—	—	1.18	—	—

Black BW A53 Schedule 40 Pipe
List Prices

Dia	List Price per Foot	
	TC Pipe	PE Pipe
1/8	$1.73	$0.85
1/4	2.08	1.15
3/8	2.39	1.34
1/2	1.35	0.80
3/4	1.61	0.95
1	2.22	1.33
1-1/4	1.97	1.73
1-1/2	2.34	2.05
2	3.17	2.77
2-1/2	5.05	4.12
3	6.61	5.40
3-1/2	8.11	6.19
4	9.56	7.27
5	—	13.63
6	—	23.37
8	—	34.74
10	—	50.02
12	—	60.26

Supplier discounts may run 20% to 50%.
Pricing for Black BW A53 Schedule 40 Grooved Pipe same as PE Pipe.

Black Malleable Standard Threaded Fittings
List Prices

Dia	90 Elbow	45 Elbow	Tee	Wye	Cross	Cap	Coupling	Brass Union
1/8	$2.17	$3.21	$3.29	—	$8.01	$1.86	$2.74	$11.29
1/4	2.17	3.21	3.10	—	6.55	1.86	2.63	10.45
3/8	2.17	3.21	3.10	7.06	6.77	1.57	2.63	7.21
1/2	1.51	2.43	2.03	6.67	7.40	1.53	2.04	6.66
3/4	1.82	3.00	2.90	8.82	9.03	2.03	2.39	7.64
1	3.17	3.80	4.94	11.86	11.12	2.50	3.65	9.98
1-1/4	5.20	6.72	8.01	20.49	18.05	3.22	4.63	14.30
1-1/2	6.84	8.30	9.97	27.85	22.37	4.42	6.26	17.84
2	11.81	12.56	16.99	43.34	36.78	6.59	9.25	61.81
2-1/2	25.77	36.33	35.88	94.05	69.54	14.70	25.61	74.40
3	38.57	47.20	52.85	152.63	96.46	21.86	34.62	—
3-1/2	91.95	95.78	108.25	—	—	43.99	—	—
4	82.70	92.54	127.75	273.94	211.27	37.45	69.71	—
5	209.62	276.48	284.67	—	—	79.29	—	—
6	249.37	332.08	368.85			145.35	—	—

Black Malleable Standard Grooved Fittings
List Prices

Dia	90 Elbow	45 Elbow	221/2 Elbow	Tee	Lateral Wye	Cross	Female Adapter	Cap	Std Flex Coupling	#150 Flange
3/4	$37	$30	$32	$57	$226	$141	$66	$31	$19	—
1	37	37	40	57	226	141	66	31	19	—
1-1/4	37	37	40	57	226	170	66	31	26	—
1-1/2	37	37	40	57	226	183	66	31	28	—
2	37	37	40	57	226	103	66	31	30	123
2-1/2	37	37	40	57	226	103	66	31	35	153
3	66	66	58	80	250	186	78	31	39	165
4	72	72	77	122	386	309	119	33	57	221
5	173	173	240	286	563	631	—	77	87	257
6	203	203	219	330	908	812	—	79	103	280
8	425	425	459	725	1,308	1,070	—	152	168	316
10	773	773	630	1,507	1,845	1,825	—	323	280	502
12	1,240	1,240	852	2,101	2,231	2,633	—	529	319	656
14	1,660	1,125	—	2,226	3,257	2,728	—	—	387	1,507
16	2,157	1,399	—	2,515	4,068	3,569	—	—	508	1,749
18	2,738	1,640	—	3,143	4,548	4,374	—	—	595	2,153
20	3,609	2,155	—	4,504	6,898	6,197	—	—	936	2,591
24	5,226	3,310	—	6,868	10,720	9,567	—	—	997	3,315

Carbon Steel Butt Weld Standard Fittings
List Prices

	List Prices Each				
	90 LR	45 LR			
Din	Elbow	Elbow	Tee	Cross	Cap
1/2	$39	$37	$95		$32
3/4	39	37	53		32
1	20	19	53	—	32
1-1/4	20	19	61	565	22
1-1/2	20	19	61	565	5
2	20	19	53	470	20
2-1/2	27	23	67	683	20
3	31	27	74	610	20
3-1/2	90	84	105	—	31
4	52	47	105	801	26
5	111	69	190	—	45
6	126	100	190	1,713	49
8	237	169	348	2,551	75
10	433	306	580	4,799	132
12	623	433	876	5,775	164
14	860	591	1,462	—	301
16	1,177	807	1,767	—	354
18	1,515	1,082	2,797	—	406
20	2,227	1,425	4,328	—	464
22	1,508	935	—	—	320
24	2,956	2,058	5,383	—	667

Carbon Steel Welding Flanges
List Prices

Dia	Flanges Raised Face #150 List Prices Each					Gasket & Bolt Sets
	Threaded	Weldneck	Slip-On	Socket	Blind	
1/2	$54.00	$10.80	$7.63	$47.00	$43.00	$3.32
3/4	54	50	35	47	43	3.32
1	54	50	35	47	43	3.32
1-1/4	54	50	35	47	43	3.32
1-1/2	54	50	35	47	43	3.32
2	54	54	35	47	43	5.30
2-1/2	66	58	62	62	69	5.60
3	66	64	49	59	55	5.87
3-1/2	33	83	49	—	22	10.66
4	77	76	61	77	75	10.66
5	113	121	110	—	102	15.82
6	127	117	99	119	117	16.58
8	215	205	150	172	191	17.72
10	391	329	261	—	367	36.97
12	549	479	386	—	554	39.14
14	—	690	512	—	946	55.45
16	—	1,175	848	—	1291	72.85
18	—	1,659	1,079	—	1625	102
20	—	2,019	1,235	—	2212	136
24	—	2,205	1,733	—	2957	182

Galvanized BW A53 Schedule 40 Pipe
List Prices

Dia	List Prices Per Foot	
	TC Pipe	PE Pipe
1/8	$2.05	$1.78
1/4	1.92	1.63
3/8	2.13	1.81
1/2	1.27	1.13
3/4	1.53	1.36
1	2.11	1.87
1-1/4	2.75	2.41
1-1/2	3.29	2.87
2	4.42	3.86
2-1/2	6.91	5.93
3	9.05	7.77
3-1/2	10.98	9.02
4	12.93	10.62

Pricing not available for Pre-Grooved Pipe.

Galvanized Malleable Standard Threaded Fittings
List Prices

Dia	List Prices Each							Brass Union
	90 Elbow	45 Elbow	Tee	Y Branch	Cross	Cap	Coupling	
1/8	$3.18	$4.69	$4.51	—	$10.52	$2.30	$3.62	$15.22
1/4	2.92	3.98	3.82	—	10.14	2.30	3.49	13.40
3/8	2.92	4.43	3.82	8.12	10.52	2.22	3.60	8.97
1/2	1.92	2.97	2.43	7.29	10.01	2.09	2.75	8.36
3/4	2.35	4.26	4.10	11.41	12.18	2.88	3.15	9.60
1	4.35	4.82	6.49	16.28	16.83	3.56	5.20	12.56
1-1/4	6.73	9.01	10.28	26.85	25.29	4.69	7.31	18.18
1-1/2	8.90	10.57	12.78	36.71	29.60	6.03	8.34	22.03
2	14.72	15.65	21.22	59.06	56.05	8.43	12.16	25.33
2-1/2	37.74	54.45	58.52	142.00	113.34	24.49	37.24	86.13
3	62.22	79.20	68.62	233.45	147.02	29.68	49.77	123.13
4	109.77	132.78	162.31	385.27	308.41	65.49	105.70	—
5	358.51	405.84	375.18	—	—	133.04	—	—
6	388.00	489.33	437.67	—	—	222.30	—	—

Copper L, K, M and ACR L-Hard Tubing
List Prices

Dia	K Hard	K Soft	L Hard	L Soft	M Hard	M Soft	Dia	ACR L Hard
1/4	$0.61	$0.72	$0.54	—	—	—	—	—
3/8	1.05	1.15	1.80	—	0.61	0.76	3/8	$0.63
1/2	1.22	1.31	1.03	1.17	0.73	0.91	1/2	0.92
5/8	—	1.64	—	—	—	1.26	5/8	1.17
3/4	2.28	2.40	1.61	1.83	1.17	1.46	3/4	1.55
1	2.90	3.08	2.25	2.57	1.69	2.11	7/8	1.83
1-1/4	3.62	3.86	3.03	3.37	2.39	2.99	1-1/8	2.61
1-1/2	4.66	4.93	3.87	4.29	3.37	4.21	1-3/8	3.50
2	7.45	8.39	6.11	7.40	5.61	7.00	1-5/8	4.54
2-1/2	10.77	—	8.92	—	7.75	9.69	2-1/8	7.30
3	14.89	—	12.08		9.85	12.31	2-5/8	10.36
3-1/2	18.20	—	—	—	—	15.15	3-1/8	14.35
4	23.21	—	19.06	—	16.65	20.81	4-1/8	23.53
5	59.65	—	53.49	—	51.19	64.00		
6	85.67	—	65.23	—	63.29	79.10		
8	160.40	—	120.37		112.27	140.34		

Wrot Copper Pressure Standard Fittings
List Prices

| Dia | Soldered Labor Hours Each | | | | | | | #125 |
	90 Elbow	45 Elbow	Tee	Adapter	Cap	Coupling	Union	Flange
1/8	$1.68	$2.17	$2.29	$4.51	$0.33	$0.28	$7.90	—
1/4	1.72	2.04	2.29	3.59	0.32	0.27	6.58	—
3/8	1.95	1.66	1.75	1.79	0.50	0.34	6.58	—
1/2	1.75	0.64	0.62	0.78	0.26	0.28	3.84	11.18
5/8	1.87	3.16	3.80	—	0.56	0.78	17.90	—
3/4	2.38	1.86	1.52	1.29	0.45	0.54	4.81	22.00
1	3.45	2.78	4.41	3.14	1.05	1.06	8.91	21.44
1-1/4	5.49	3.77	6.67	4.63	1.49	1.87	—	26.65
1-1/2	7.36	4.52	9.26	5.32	2.18	2.46	—	43.36
2	19.41	7.58	14.42	9.00	3.99	4.09	—	54.56
2-1/2	30.72	16.14	29.13	28.71	12.79	8.72	—	70.40
3	43.24	23.96	44.44	39.15	17.53	13.88	—	70.40
3-1/2	165	45	128	—	53	25	—	—
4	113	51	107	65	33	26	—	10
5	206	198	334	—	—	61	—	204
6	287	296	458	—	—	95	—	224
8	1159	1059	1857	—	—	329	—	561

PVC Schedule 40 Pipe and Standard Fittings
List Prices

Dia	Pipe Per Ft	90 D Elbow	45 D Elbow	30 D Elbow	22-1/2 D Elbow	15 D Elbow	11-1/4 D Elbow	Tee	Wye	Slip Cross
1/8	$0.61	—		—	—	—		—	—	—
1/4	0.89	—		—	—	—		—	—	—
3/8	1.17	$0.71	—	—	—	—	—	$1.15	—	—
1/2	0.67	0.30	$1.96	—	—	—	—	0.37	—	$4.51
3/4	0.90	1.33	3.02	—	—	—	—	0.42	—	7.57
1	1.32	2.41	3.63	—	—	—	—	0.79	—	9.39
1-1/4	1.79	4.24	5.14	—	—	—	—	1.25	—	12.44
1-1/2	2.10	4.51	6.40	—	—	—	—	1.51	—	14.09
2	2.82	7.14	8.35	$13.38	$8.31	$13.88	$13.38	2.20	—	20.77
2-1/2	4.57	21.65	21.65	—	—	—	—	7.25	—	43.37
3	5.82	25.91	33.66	16.87	19.93	16.87	17.91	38.02	$27.92	53.21
3-1/2	10.57	—	—	—	—	—	—	—	—	—
4	8.28	46.39	60.46	20.23	24.00	20.33	22.85	67.80	52.65	78.84
5	11.59	119.93	119.93	—	—	—	—	163.83	—	—
6	14.59	147.53	149.28	32.64	32.64	32.64	37.64	228.35	111.41	301.19
8	21.90	379.95	359.14	55.19	55.19	55.19	55.19	529.59	319.81	387.95
10	31.63	357.75	233.71	83.28	83.28	83.28	83.28	445.35	532.64	469.56
12	41.79	479.80	346.35	148.71	131.75	131.75	131.75	658.02	724.08	661.72
14	57.32	480.60	319.25	207.13	207.13	207.13	207.13	772.89	1025.83	1176.65
16	74.82	687.72	421.93	282.36	282.36	282.36	282.36	1118.81	1326.01	1491.66

PVC Schedule 40 Standard Fittings
List Prices

	List Prices Each								
Dia	Adapter SxF	Adapter SxM	Slip Cap	Female Thrd Cap	Male Thrd Plug	Fitting Plug	Coupling SxS	Union SxS	Slip-on Flange
1/4	—	—	—	—	—	—	—	—	—
3/8	$1.09	$1.02	$0.76	$1.03	$0.83	$0.80	$0.53	—	—
1/2	1.33	1.08	0.27	2.28	3.19	2.57	0.80	$2.91	$3.12
3/4	1.69	1.19	0.31	2.57	3.45	3.02	1.08	3.33	3.35
1	1.96	2.11	0.49	3.92	5.62	3.64	1.87	3.41	3.73
1-1/4	3.02	2.56	0.69	4.68	5.92	4.88	2.56	11.00	3.84
1-1/2	3.48	3.44	0.76	4.88	6.38	6.22	2.75	11.47	3.92
2	4.68	4.55	0.91	8.63	8.19	7.73	4.28	15.46	5.21
2-1/2	11.85	13.49	2.91	12.75	11.37	12.51	9.39	—	8.05
3	15.93	19.72	3.19	16.69	17.28	17.42	14.74	—	8.86
4	26.37	25.15	7.25	29.94	38.97	39.40	21.27	—	11.23
5	68.23	41.46	12.17	56.62	—	—	38.99	—	—
6	96.98	65.64	17.36	69.13		—	67.33	—	17.66
8	182.83	74.56	43.62	—		—	125.68	—	28.62
10	—	—				—	4.31	—	—
12						—	65.22	—	

Pipe Hangers
List Prices Each

Size	Band Hours	Band Price	Adjustable Swivel Roller Hours	Adjustable Swivel Roller Price	Clevis Hours	Clevis Price	Riser Clamp Hours	Riser Clamp Price	Weld Beam Attachment Hours	Weld Beam Attachment Price
3/4	0.60	$1.68			0.60	$1.79	0.13	$5.58	—	$15.40
1	0.60	1.95	—	—	0.60	1.90	0.13	5.63	—	31.85
1-1/4	0.60	2.00	0.60	$30.64	0.60	2.16	0.13	6.84	—	84.20
1-1/2	0.60	2.26	0.60	31.17	0.60	2.37	0.13	7.21	—	145.30
2	0.60	3.11	0.60	31.75	0.60	2.63	0.13	7.58	—	222.30
2-1/2	0.64	3.26	0.64	37.49	0.64	4.26	0.15	8.00	—	287.75
3	0.64	3.58	0.64	38.96	0.64	5.37	0.15	8.74	—	399.50
4	0.69	5.58	0.69	52.91	0.69	6.48	0.16	11.00	—	—
5	0.69	5.90	0.69	61.71	0.69	8.95	0.16	15.22	—	—
6	0.73	8.16	0.73	75.45	0.73	10.32	0.16	28.80	—	—
8	0.77	12.74	0.77	115.62	0.77	17.01	0.17	29.64	—	—
10	—	—	0.77	135.10	0.77	32.17	0.17	44.23	—	—
12	—		0.77	162.58	0.77	38.22	0.20	52.33	—	—
14	—		0.86	258.62	0.86	55.65	0.20	89.98	—	—
16	—	—	0.86	314.43	0.86	79.29	0.22	207.39	—	—
18	—	—	0.97	340.86	0.97	98.46	0.22	228.34	—	—
20	—	—	1.07	494.65	1.07	206.91	0.22	264.36	—	—
24	—	—	1.18	783.54	1.18	365.50	—	—	—	—
30	—	—	1.29	1878.55	1.29	473.01	—	—	—	—

Threaded Rods, Concrete Inserts and Anchors

Size	Threaded Hrs/Ft	Threaded Price/Ft	Concrete Insert Hrs/Ea	Concrete Insert Price/Ea	Concrete Anchor Hrs/Ea	Concrete Anchor Price/Ea
1/4	0.052	$0.33	0.080	$3.45	—	—
3/8	0.052	0.51	0.080	3.59	—	$2.05
1/2	0.052	0.99	0.080	3.59	—	3.75
5/8	0.052	1.56	0.080	3.78	—	6.95
3/4	0.	2.53	0.080	4.00	—	12.25
7/8	0.052	3.37	0.080	4.68	—	—

Typical Hanger Spacing
Feet Between Supports

Pipe Size	Steel	Copper	Grooved Piping	Cast Iron & No Hub	Plastic (Rigid)
1/2	8	6	6	—	3
3/4	8	8	8	—	3
1	8	8	8	—	3-1/2
1-1/4	10	10	10	—	3-1/2
1-1/2	10	10	10	5	3-1/2
2	10	10	10	5	4
2-1/2	11	12	12	—	4
3	12	12	12	5	4-1/2
4	14	12	12	5	4-1/2
5	15	12	14	5	4-1/2
6	16	14	14	5	5
8	18	14	14	5	5
10	21	—	16	5	5-1/2
12	22	—	16	5	5-1/2
14	25	—	18	5	—
16	27	—	18	—	—
18	28	—	20	—	—
20	30	—	20	—	—
24	32	—	20	—	—

Pipe Sleeves and Cut Openings
Labor and Material

Material	Pipe		Sheet Metal		Pre Fab		Wood Box		Cut Hole	
Size	Hours	Price	Hours	Price	Hours	Price	Hours	Price	Hours	Price
2	0.56	$3.45	0.30	—	0.25	—	0.50	—	0.50	$1.95
3	0.56	7.07	0.30	$0.85	0.35	—	0.65	—	0.50	4.32
4	0.75	10.22	0.40	0.85	0.40	—	0.65	—	0.60	4.32
6	0.85	13.54	0.50	1.50	0.45	—	0.70	—	0.80	7.11
8	0.95	18.16	0.60	2.00	0.45	—	0.90	—	1.00	9.72
10	1.00	24.84	0.70	3.35	0.50	—	1.40	—	1.40	13.01
12	1.20	28.84	0.80	4.00	0.50	—	1.50	—	1.50	15.70
14	1.30	—	0.80	4.75	0.60	—	1.70	—	2.40	19.08
16	1.50		0.90	5.20	0.65	—	1.90	—	3.00	22.06
18	1.60		1.00	5.90	0.70	—	2.00	—	3.50	24.72

Hanger Rod Sizes and Weights

Pipe Sizes	Rod Size	Weight Lbs/Ft
3/4-2	3/8	0.376
2-1/2 - 3-1/2	1/2	0.668
4-5	5/8	1.045
6	3/4	1.500
8-12	7/8	2.045
14-16	1	2.670
18	1-1/8	3.380
20	1-1/4	4.175
24	1-1/2	6.000
30	1-7/8	9.338

Fiberglass Blanket ASJ 3Lb Insulation
Labor

Dia	Labor Hours per Foot			
	Insulation Thickness			
	1/2″	1″	1-1/2″	2″
1/8	—	—	—	—
1/4	—	—	—	—
3/8	—	—	—	—
1/2	0.05	0.06	0.06	0.06
5/8	0.05	0.06	0.06	0.06
3/4	0.06	0.06	0.06	0.07
7/8	0.06	0.06	0.07	0.07
1	0.06	0.06	0.07	0.07
1-1/4	0.07	0.07	0.07	0.08
1-1/2	0.07	0.07	0.08	0.08
2	0.07	0.08	0.08	0.09
2-1/2	0.07	0.08	0.09	0.09
3	0.08	0.08	0.09	0.10
3-1/2	0.08	0.09	0.10	0.10
4	0.09	0.09	0.10	0.11
5	0.10	0.10	0.11	0.12
6	0.11	0.12	0.13	0.15
8	0.12	0.15	0.17	0.19
10	—	0.17	0.19	0.21
12	—	0.19	0.20	0.23
14	—	0.19	0.22	0.25
16	—	0.21	0.24	0.27
18	—	0.21	0.25	0.30
20	—	0.25	0.28	0.34
24	—	0.25	0.30	0.38

Fiberglass Preform ASJ 3Lb Insulation
Labor

| Dia | Labor Hours per Foot | | | |
| | Insulation Thickness | | | |
	1/2″	1″	1-1/2″	2″
1/8	0.04	—	—	—
1/4	0.04	—	—	—
3/8	0.05	—	—	—
1/2	0.05	0.05	0.05	0.06
5/8	0.05	0.05	0.06	0.06
3/4	0.06	0.06	0.06	0.06
7/8	0.06	0.06	0.06	0.07
1	0.06	0.06	0.07	0.07
1-1/4	0.07	0.07	0.07	0.07
1-1/2	0.07	0.07	0.07	0.08
2	0.07	0.07	0.08	0.08
2-1/2	0.07	0.08	0.08	0.09
3	0.08	0.08	0.09	0.10
3-1/2	0.08	0.09	0.10	0.11
4	0.09	0.10	0.11	0.12
5	0.10	0.11	0.12	0.13
6	0.11	0.12	0.14	0.15
8	0.12	0.14	0.16	0.17
10	0.15	0.18	0.18	0.20
12	0.19	0.20	0.22	0.23
14	—	0.23	0.25	0.26
16	—	0.27	0.29	0.30
18	—	0.30	0.32	0.34
20	—	0.36	0.37	0.39
24	—	0.40	0.42	0.43

Foam Plastic Insulation
Labor

| Dia | Labor Hours per Foot | | | |
| | Insulation Thickness | | | |
	3/8"	1/2"	3/4"	1"
1/8	0.05	0.06	0.06	0.07
1/4	0.05	0.06	0.06	0.07
3/8	0.05	0.06	0.06	0.07
1/2	0.05	0.06	0.07	0.07
5/8	0.05	0.06	0.07	0.07
3/4	0.06	0.06	0.07	0.07
7/8	0.06	0.06	0.07	0.07
1	0.06	0.06	0.07	0.07
1-1/4	0.06	0.06	0.07	0.08
1-1/2	0.06	0.06	0.07	0.08
2	0.06	0.06	0.07	0.08
2-1/2	0.06	0.07	0.07	0.08
3	0.06	0.07	0.08	0.09
3-1/2	0.07	0.07	0.08	—
4	0.07	0.07	0.08	

Calcium Silicate Insulation
Labor

| Dia | Labor Hours per Foot | | | |
| | Insulation Thickness | | | |
	1"	1-1/2"	2"	3"
1/2	0.08	0.08	0.09	0.12
5/8	0.08	0.08	0.09	0.12
3/4	0.08	0.09	0.09	—
7/8	0.08	0.09	0.09	—
1	0.08	0.09	0.10	0.13
1-1/4	0.08	0.09	0.10	0.13
1-1/2	0.08	0.09	0.10	0.13
2	0.09	0.09	0.10	0.14
2-1/2	0.09	0.10	0.11	0.14
3	0.09	0.10	0.11	0.15
3-1/2	0.10	0.10	0.11	0.15
4	0.10	0.10	0.12	0.16
5	0.10	0.11	0.12	0.16
6	0.11	0.11	0.13	0.17
8	—	0.14	0.14	0.18
10	—	0.15	0.16	0.19
12	—	0.16	0.17	0.22
14	—	0.17	0.18	0.23
16	—	0.18	0.19	0.25
18	—	0.20	0.20	0.27
20	—	—	0.23	0.30
22	—		0.25	0.34
24	—		0.27	0.37

Rigid Eurethane Insulation
Labor

Dia	Labor Hours per Foot			
	Insulation Thickness			
	1/2″	1″	1-1/2″	2″
1/8	0.04	0.05	0.05	0.05
1/4	0.04	0.05	0.05	0.05
3/8	0.05	0.05	0.05	0.06
1/2	0.05	0.05	0.06	0.06
5/8	0.06	0.06	0.06	0.06
3/4	0.06	0.06	0.06	0.06
7/8	0.06	0.06	0.06	0.06
1	0.06	0.06	0.06	0.06
1-1/4	0.06	0.06	0.06	0.07
1-1/2	0.06	0.06	0.07	0.07
2	0.06	0.07	0.07	0.07
2-1/2	0.07	0.07	0.07	0.08
3	0.08	0.07	0.08	0.09
3-1/2	0.08	0.09	0.10	0.11
4	0.09	0.10	0.11	0.12
5	0.10	0.11	0.12	0.13
6	0.11	0.12	0.13	0.14
8	0.14	0.15	0.16	0.17
10	0.17	0.19	0.20	0.19
12	0.20	0.21	0.22	0.24

Foam Rubber Tubing Insulation
Labor

| Dia | Labor Hours per Foot | | |
| | Insulation Thickness | | |
	3/8″	1/2″	3/4″
1/4	0.06	0.07	0.07
3/8	0.06	0.07	0.07
1/2	0.06	0.07	0.07
5/8	0.06	0.07	0.07
3/4	0.06	0.07	0.07
7/8	0.06	0.07	0.08
1	0.07	0.08	0.08
1-1/4	0.07	0.08	0.08
1-1/2	0.07	0.08	0.08
2	0.07	0.08	0.08
2-1/2	0.07	0.08	0.09
3	0.08	0.09	0.09
3-1/2	0.08	0.09	0.09
4	0.08	0.09	0.09
5	0.08	0.09	0.10
6	0.08	0.10	0.10

Fiberglass Preform ASJ 3Lb Insulation
Installed Prices with Overhead and Profit

| Dia | Labor Hours per Foot | | | |
| | Insulation Thickness | | | |
	1/2"	1"	1-1/2"	2"
1/2	$3.04	$3.29	$4.47	$5.77
3/4	3.23	3.54	4.65	6.00
1	3.38	3.69	4.89	6.38
1-1/4	3.59	3.94	5.21	6.75
1-1/2	3.69	4.06	5.35	6.93
2	3.89	4.31	5.72	7.26
2-1/2	4.20	4.64	6.14	7.77
3	4.52	5.03	6.42	8.33
3-1/2	4.79	5.40	6.93	8.93
4	5.54	6.19	7.58	9.86
5	6.05	6.89	8.33	10.89
6	7.07	7.68	9.27	12.00
8	—	9.91	11.40	14.88
10	—	11.30	13.49	17.44
12	—	12.79	14.56	18.60
14	—	13.77	16.23	21.39
16	—	16.05	17.86	23.25
18	—	16.65	19.53	25.11
20	—	18.60	21.39	27.90
24	—	21.39	24.18	31.62

Foam Plastic Insulation
Installed Prices with Overhead and Profit

Dia	Labor Hours per Foot			
	Insulation Thickness			
	3/8″	1/2″	3/4″	1″
1/8	$2.41	$2.71	$3.02	—
1/4	2.44	2.72	3.04	$3.46
3/8	2.46	2.73	3.11	3.55
1/2	2.55	2.78	3.26	3.82
3/4	2.64	2.87	3.37	3.94
1	2.74	2.99	3.57	4.34
1-1/4	2.85	3.13	3.85	4.53
1-1/2	2.90	3.26	3.95	4.58
2	3.04	3.44	4.42	5.03
2-1/2	3.27	3.70	4.74	5.30
3	3.39	4.01	5.07	5.58
4	—	4.49	5.77	—

Calcium Silicate Insulation
Installed Prices with Overhead and Profit

| | Labor Hours per Foot | | | |
| | Insulation Thickness | | | |
Dia	1/2″	1″	2″	3″
1/2	$5.21	$5.82	$7.49	—
3/4	5.21	5.86	7.63	—
1	5.12	6.05	7.82	—
1-1/4	5.26	6.38	8.28	$10.98
1-1/2	5.26	6.51	8.47	11.30
2	5.77	6.89	8.84	11.96
2-1/2	5.96	7.21	9.77	12.79
3	6.33	7.49	9.72	12.84
3-1/2	7.03	—	—	—
4	7.49	8.47	10.79	15.35
5	7.96	9.07	11.86	—
6	5.82	9.63	12.98	17.72
8	—	11.68	17.44	20.46
10	—	14.28	20.46	25.11
12	—	17.17	23.25	26.04
14	—	21.39	25.11	29.76
16	—	24.18	26.97	32.55
18	—	26.04	29.76	35.34

Foam Rubber Tubing Insulation
Installed Prices with Overhead and Profit

| | Labor Hours per Foot | | |
| | Insulation Thickness | | |
Dia	3/8"	1/2"	3/4"
1/4	$2.72	$3.67	$3.85
3/8	2.75	3.72	3.93
1/2	2.89	3.79	4.07
3/4	2.94	3.85	4.23
1	3.10	3.93	4.39
1-1/4	3.19	4.05	4.79
1-1/2	3.29	4.18	4.93
2	3.69	4.39	5.26
2-1/2	4.17	4.65	5.86
3	4.30	5.21	6.42
3-1/2	4.79	5.26	6.70
4	—	5.63	7.12
5	—	6.51	7.82
6	—	—	10.56

Chapter 18

Valves and Specialties

This chapter covers labor and price tables for valves and specialties in pressure type piping systems as follows:

Standard Valves:
- Bronze Valves, Threaded and Soldered, 125#, 150#, 300#
- Iron Valves, Flanged and Welded, 125#, 150#, 250#
- Ductile Iron Valves, Flanged and Grooved, #200
- Plastic PVC Valves, Socket, Threaded and Flanged
- Carbon Steel Valves, Threaded and Welded, #250, #1000

Specialty Valves:
- Strainers
- Control Valves
- Steam Traps
- Pressure Valves
- Gas Valves
- Temperature Control Valves
- Balancing Valves
- Gas Valves

Specialties:
- Hot Water
- Refrigeration
- Chilled Water
- Steam

Units

All **diameters** are in inches.
Flanges and hangers are covered in chapter 17.

Labor

The labor productivity rates in the tables are what the average mechanic under average conditions can erect on lower floors of construction projects. However, they may require the application of **labor correction** factors for variable conditions such as upper floors, higher piping runs, congested spaces, etc.

The labor in the tables are in hours per piece and hours per foot for pipe.

Pricing and Discounts

The pricing in the tables are based on the typical manufacturers list prices and the contractor's supplier discounts must be applied.

Appropriate price discounts can range from roughly 10 percent to 80 percent dependent on the following:

- Supplier discount structure
- Who manufacturer is
- Size of batch involved
- Availability
- Shipping distance, costs

Not All List Prices Are Publicly Available

Where there are no prices listed in tables, they are not usually publicly available from the manufacturer. If there is neither labor nor prices listed for a particular diameter in the tables, the particular item may not be available or it may not be a commonly used item.

Caution With List Prices

Pricing can sometimes vary significantly because of changes in commodity material pricing such as copper, etc. Periodic checks of current list prices should be made.

If in doubt about your discounts from list prices, *check with your supplier* as to what they are for the various categories of items - or request a *quotation*.

Automatic Updating Available in Wendes Computer System

Automatic updating of list pricing is available in the *Wendes Computerized Piping Estimating System*. Call (847) 818-8371 for more information.

Bronze #125 Valves
Threaded and Soldered Connections

Diameter	Threaded		Soldered	
	Labor	Price	Labor	Price
Angle Valve				
1/4	0.37	$33.40	0.29	$22.80
3/8	0.37	33.40	0.29	22.80
1/2	0.37	33.40	0.29	22.80
3/4	0.50	45.60	0.39	31.30
1	0.61	59.60	0.48	41.00
1-1/4	0.71	88.60	0.54	—
1-1/2	0.78	103.40	0.58	153.00
2	0.96	155.90	0.69	222.00
2-1/2	1.53	279.90	1.11	—
3	1.88	592.20	1.38	—
Ball Valve				
1/2	0.36	$5.50	0.34	$5.50
3/4	0.50	7.70	0.45	7.70
1	0.61	12.65	0.55	12.65
1-1/4	0.71	22.10	0.60	22.10
1-1/2	0.79	27.80	0.65	27.80
2	0.95	35.10	0.73	35.10
Check Valve				
3/8	0.37	$14.80	0.29	$14.80
1/2	0.37	15.80	0.29	15.80
3/4	0.50	19.50	0.39	19.50
1	0.61	25.90	0.48	25.90
1-1/4	0.71	32.70	0.54	32.70
1-1/2	0.78	43.30	0.58	43.30
2	0.96	60.00	0.69	60.00
2-1/2	1.53	—	—	—
3	1.88	—	—	—

Bronze #125 Valves
Threaded and Soldered Connections

Diameter	Threaded		Soldered	
	Labor	Price	Labor	Price
Gate Valve				
1/4	0.36	$24.00	0.29	—
3/8	0.36	24.53	0.29	$22.00
1/2	0.36	21.16	0.29	21.37
3/4	0.50	26.11	0.39	24.63
1	0.61	33.26	0.46	34.84
1-1/4	0.73	49.37	0.52	52.53
1-1/2	0.81	57.68	0.57	59.37
2	0.97	75.68	0.66	83.37
2-1/2	1.56	198.11	1.05	194.95
3	1.89	241.89	1.29	274.74
Globe Valve				
1/8	0.38	$33.40	—	—
1/4	0.38	33.40	—	—
3/8	0.38	35.26	0.29	$36.00
1/2	0.38	35.26	0.29	36.00
3/4	0.50	45.68	0.39	45.89
1	0.62	62.11	0.48	62.63
1-1/4	0.73	82.32	0.54	81.37
1-1/2	0.82	106.95	0.58	106.00
2	1.02	162.00	0.69	160.42
2-1/2	1.62	370.00	1.11	373.00
3	1.99	527.00	1.38	532.00

Bronze #150 Valves
Threaded and Soldered Connections

Diameter	Threaded		Soldered	
	Labor	Price	Labor	Price
Angle Valve				
1/8	0.37	$44.70	—	—
1/4	0.37	54.46	—	—
3/8	0.37	54.46	—	—
1/2	0.37	54.46	—	—
3/4	0.50	74.01	—	—
1	0.61	106.93	—	—
1-1/4	0.71	139.51	—	—
1-1/2	0.78	186.39	—	—
2	0.96	292.51	—	—
2-1/2	1.53	478.30	—	—
3	1.88	750.00	—	—
Ball Valve				
1/4	0.36	$9.40	0.34	$9.40
3/8	0.36	9.40	0.34	9.40
1/2	0.36	9.40	0.34	9.40
3/4	0.50	15.30	0.45	15.30
1	0.61	19.40	0.55	19.40
1-1/4	0.71	32.60	0.60	32.60
1-1/2	0.79	41.60	0.65	41.60
2	0.95	52.00	0.73	52.00
2-1/2	1.61	245.00	1.27	245.00
3	1.90	273.00	1.50	273.00
Check Valve				
1/4	0.37	$40.10	0.29	$41.20
3/8	0.37	40.10	0.29	41.20
1/2	0.37	40.10	0.29	41.20
3/4	0.50	47.50	0.39	49.70
1	0.61	64.00	0.48	68.00
1-1/4	0.71	86.00	0.54	88.00
1-1/2	0.78	116.00	0.58	117.00
2	0.96	174.00	0.69	171.00
2-1/2	1.53	250.00	1.11	272.00
3	1.88	371.00	1.38	404.00

Bronze #150 Valves
Threaded and Soldered Connections

Diameter	Threaded		Soldered	
	Labor	Price	Labor	Price
Gate Valve				
1/4	0.36	$41.00	—	—
3/8	0.36	41,00	0.29	$49.60
1/2	0.36	39.50	0.29	47.50
3/4	0.50	43.50	0.39	58.00
1	0.61	53.10	0.46	76.00
1-1/4	0.73	73.90	0.52	108.00
1-1/2	0.81	92.50	0.57	131.00
2	0.97	125.40	0.66	189.00
2-1/2	1.56	337.60	1.05	389.00
3	1.89	464.50	1.29	530.00
3-1/2	2.29	944.00	—	—
4	2.29	1145.00	1.85	1114.00
5	—	—	2.22	—
6	—	—	2.50	—
Globe Valve				
1/4	0.38	$43.37	0.29	$53.00
3/8	0.38	43.37	0.29	$53.00
1/2	0.38	43.37	0.29	52.00
3/4	0.50	58.63	0.39	69.00
1	0.62	91.26	0.48	98.00
1-1/4	0.73	145.05	0.54	153.00
1-1/2	0.82	173.05	0.58	186.00
2	1.02	261.37	0.69	275.00
2-1/2	1.62	535.58	1.11	538.00
3	1.99	762.00	1.38	693.00

Bronze #300 Valves
Threaded and Soldered Connections

Diameter	Threaded		Soldered	
	Labor	Price	Labor	Price
Angle Valve				
1/4	0.37	$83.79	—	—
3/8	0.37	83.79	—	—
1/2	0.37	63.68	—	—
3/4	0.50	101.37	—	—
1	0.61	170.42	—	—
1-1/4	0.71	259.16	—	—
1-1/2	0.78	259.16	—	—
2	0.96	323.47	—	—
2-1/2	1.53	773.00	—	—
3	1.88	1062.00	—	—
Ball Valve				
1/2	0.36	7.50	—	—
3/4	0.50	9.26	—	—
1	0.61	13.61	—	—
1-1/4	0.71	23.15	—	—
1-1/2	0.79	29.99	—	—
2	0.95	42.72	—	—
Check Valve				
1/4	—	—	0.29	—
3/8	0.37	$41.70	0.29	—
1/2	0.37	41.70	0.29	—
3/4	0.50	52.00	0.39	—
1	0.61	70.00	0.48	—
1-1/4	0.71	88.00	0.54	—
1-1/2	0.78	119.00	0.58	—
2	0.96	177.00	0.69	—
2-1/2	—	—	1.11	—
3	—	—	1.38	—

Bronze #300 Valves
Threaded and Soldered Connections

Diameter	Threaded		Soldered	
	Labor	Price	Labor	Price
Gate Valve				
1/4	0.38	$67.00	—	—
3/8	0.38	67.00	—	—
1/2	0.38	65.00	—	—
3/4	0.52	169.80	—	—
1	0.65	191.90	—	—
1-1/4	0.76	225.00	—	—
1-1/2	0.86	235.10	—	—
2	1.11	345.50	—	—
Globe Valve				
1/4	0.37	$83.88	—	—
3/8	0.37	85.88	—	—
1/2	0.37	89.24	—	—
3/4	0.52	127.07	—	—
1	0.70	180.58	—	—
1-1/4	0.76	255.38	—	—
1-1/2	0.86	321.74	—	—
2	1.04	429.23	—	—
2-1/2	1.76	649.00	—	—
3	2.18	1030.00	—	—

Iron #125 Valves
Threaded and Flanged Connections

Diameter	Threaded		Flanged	
	Labor	Price	Labor	Price
Angle Valve				
2	—	—	0.85	$792.00
2-1/2	—	—	0.94	850.00
3	—	—	1.05	924.00
4	—	—	1.90	1315.00
5	—	—	2.39	2568.00
6	—	—	2.75	2568.00
8	—	—	3.93	4595.00
Check Valve, Swing				
2	0.96	—	0.85	$205.70
2-1/2	1.53	—	0.94	263.40
3	1.88	—	1.05	2.84
4	2.01	—	1.90	447.40
5	2.53	—	2.39	774.80
6	2.81	—	2.75	762.20
8	—	—	3.93	1434.10
10	—	—	5.47	2447.00
12	—	—	6.42	3804.50
14	—	—	8.76	10180.50
16	—	—	9.61	13042.40
18	—	—	10.34	17355.90

Iron #125 Valves
Threaded and Flanged Connections

Diameter	Threaded		Flanged	
	Labor	Price	Labor	Price
Gate Valve, Iron				
2	0.97	$339.00	0.99	$344.60
2-1/2	1.56	377.00	1.11	385.20
3	1.89	411.00	1.21	385.20
4	2.26	570.00	2.26	568.20
5	—	—	2.79	939.00
6	—	—	3.04	953.00
8	—	—	4.17	1756.10
10	—	—	5.78	3644.80
12	—	—	6.62	5278.30
Globe Valve, Bronze Mounted				
2	—	—	0.94	$585.20
2-1/2	—	—	1.11	619.50
3	—	—	1.25	716.30
4	—	—	2.20	1024.80
5	—	—	2.83	1868.80
6	—	—	3.18	1834.60
8	—	—	4.43	3657.10
10	—	—	6.22	6425.70

Iron #150 Valves
Threaded and Flanged Connections

Diameter	Threaded		Flanged	
	Labor	Price	Labor	Price
Gate Valve, RS U-Bolt Bonnet				
1/4	0.36	$78.00	—	—
3/8	0.36	74.00	—	—
1/2	0.36	70.00	—	—
3/4	0.50	70.00	—	—
1	0.61	70.00	—	—
1-1/4	0.73	92.00	—	—
1-1/2	0.81	98.00	—	—
2	0.97	142.00	0.99	—
2-1/2	1.56	213.00	1.11	—
3	1.89	277.00	1.21	—
4	2.29	642.00	2.26	—
Check Valve, Wafer				
1-1/2	—	—	0.64	$131.00
2	—	—	0.64	175.00
2-1/2	—	—	0.64	192.00
3	—	—	0.64	206.00
4	—	—	1.29	263.00
5	—	—	1.52	344.00
6	—	—	1.52	460.00
8	—	—	1.76	822.00
10	—	—	2.93	1422.00
12	—	—	3.34	2715.00

Iron #150 Valves
Threaded and Flanged Connections

Diameter	Threaded		Flanged	
	Labor	Price	Labor	Price
Butterfly Valve, Wafer				
2	—	—	0.66	$130.75
2-1/2	—	—	0.66	105.05
3	—	—	0.66	129.63
4	—	—	1.31	155.53
5	—	—	1.52	210.09
6	—	—	1.52	421.30
8	—	—	1.61	—
10	—	—	2.56	—
12	—	—	2.76	—
14	—	—	4.02	—
16	—	—	5.14	—
18	—	—	6.08	—
20	—	—	7.31	—
24	—	—	8.92	

Iron #250 Valves
Threaded and Flanged Connections

Diameter	Threaded		Flanged	
	Labor	Price	Labor	Price
Angle Valve				
2-1/2	—	—	1.88	$1,976.00
3	—	—	2.08	2254.00
4	—	—	2.65	2833.00
6	—	—	4.24	4809.00
8	—	—	5.82	8404.00
Check Valve, Wafer				
2	—	—	1.03	$226.00
2-1/2	—	—	1.29	270.00
3	—	—	1.35	305.00
4	—	—	1.37	409.00
5	—	—	1.41	552.00
6	—	—	2.11	705.00
8	—	—	2.44	1415.00
10	—	—	3.78	2092.00

Iron #250 Valves
Threaded and Flanged Connections

Diameter	Threaded		Flanged	
	Labor	Price	Labor	Price
Gate Valve, OS & Y				
2	—	—	1.56	$672.60
2-1/2	—	—	2.02	820.90
3	—	—	2.53	930.60
4	—	—	3.29	1349.60
5	—	—	4.10	2284.20
6	—	—	5.41	2284.20
8	—	—	6.41	3959.60
10	—	—	8.54	5922.70
12	—	—	21.14	9243.10
Globe Valve				
2	—	—	1.42	$734.00
2-1/2	—	—	2.06	1057.00
3	—	—	2.29	1095.00
4	—	—	2.89	1601.00
6	—	—	5.03	2877.00
8	—	—	5.99	4880.00

Ductile Iron #200 Valves
Flanged and Grooved Connections

Diameter	Lug Style		Wafer Style	
	Labor	Price	Labor	Price
Butterfly Valve, Flanged				
2	0.66	$100.00	0.10	$84.00
2-1/2	0.66	102.00	0.10	$89.00
3	0.66	114.00	0.14	96.00
4	1.32	143.00	0.20	120.00
5	1.53	207.00	0.23	171.00
6	1.61	244.00	0.31	209.00
8	1.77	338.00	0.48	314.00
10	2.69	475.00	0.71	434.00
12	3.00	751.00	1.24	668.00

Grooved Connection

Diameter	Man Handle		Gear Handle	
	Labor	Price	Labor	Price
Butterfly Valve, Grooved				
1-1/2	—	—	0.66	$745.30
2	0.63	$312.75	0.76	661.85
2-1/2	0.62	396.60	0.76	745.40
3	0.72	448.50	0.87	764.95
4	1.05	493.85	1.05	856.10
5	1.52	942.35	1.52	1291.70
6	1.83	1081.70	1.83	1371.15

Plastic Valves
Socket, Threaded and Flanged Connections

Diameter	Socket		Threaded		Flanged	
	Labor	Price	Labor	Price	Labor	Price
Ball Valve, True Union Teflon Seat						
1/2	0.28	$30.00	0.38	$30.00	0.71	$74.92
3/4	0.38	35.51	0.50'	35.51	0.71	79.87
1	0.45	42.33	0.61	42.33	0.71	93.61
1-1/4	0.50	70.28	0.72	70.28	0.76	126.90
1-1/2	0.56	70.28	0.80	70.28	0.82	126.90
2	0.65	92.05	0.99	92.05	0.87	134.30
3	—	—	—	—	1.11	303.90
4	—	—	—	—	2.22	524.00
Ball Valve True Union Multi-Port						
1/2	0.28	$64.91	0.38	$64.91	0.71	—
3/4	0.38	67.91	0.50	67.91	0.71	—
1	0.45	73.14	0.61	73.14	0.71	—
1-1/4	0.50	116.70	0.72	116.70	0.76	—
1-1/2	0.56	116.70	0.80	116.70	0.82	—
2	0.65	154.70	0.99	154.70	0.87	—
Union Ball Valve, Teflon Seat						
1/2	0.28	$30.00	0.38	$27.08	—	—
3/4	0.38	35.51	0.50	35.51	—	—
1	0.45	42.33	0.61	42.33	—	—
1-1/4	0.50	70.28	0.72	70.28	—	—
1-1/2	0.56	70.28	0.80	70.28	—	—
2	0.65	92.05	0.99	90.25	—	—

Plastic Valves
Socket, Threaded and Flanged Connections

Diameter	Socket		Threaded		Flanged	
	Labor	Price	Labor	Price	Labor	Price
Ball Check Valve						
1/2	0.28	$31.34	0.38	$31.34	0.71	—
3/4	0.38	37.30	0.50	37.30	0.71	—
1	0.45	47.69	0.61	47.69	0.71	—
1-1/4	0.50	—	0.72	—	0.76	—
1-1/2	0.56	93.25	0.80	93.25	0.82	—
2	0.65	114.65	0.99	114.65	0.87	$661.29
3	—	—	1.92	246.11	1.11	1104.30
Butterfly Valve						
4	—	—	—	—	0.34	$5.50
6	—	—	—	—	0.45	7.70
8	—	—	—	—	0.55	12.65
10	—	—	—	—	0.60	22.10

Cast Iron #250 Valves
Threaded and Welded Connections

Diameter	Threaded		Welded	
	Labor	Price	Labor	Price
Steam Ball Valve, Double Union				
1/4	0.36	—	—	—
3/8	0.36	—	—	—
1/2	0.36	$69.53	0.71	—
3/4	0.49	98.36	0.85	—
1	0.61	122.66	1.08	—
1-1/4	0.69	172.42	1.25	—
1-1/2	0.77	226.81	1.51	—
2	0.94	260.37	1.93	—
Steam Ball Valve				
1/4	0.36	$21.37	—	—
3/8	0.36	$21.37	—	—
1/2	0.36	$22.61	—	—
3/4	0.49	35.40	—	—
1	0.61	43.20	—	—
1-1/4	0.69	67.12	—	—
1-1/2	0.77	85.64	—	—
2	0.94	112.25	—	—

Cast Iron #250 Valves
Threaded and Welded Connections

Diameter	Threaded		Welded	
	Labor	Price	Labor	Price
Steam Ball Valve, High Pressure				
1/2	0.36	$42.69	—	—
3/4	0.49	54.55	—	—
1	0.61	77.33	—	—
1-1/4	0.69	111.83	—	—
1-1/2	0.77	151.56	—	—
2	0.94	177.51	—	—
Steam Ball Valve, 3 pc Full Port, High Pressure				
1/4	0.36	$61.79	—	—
3/8	0.36	61.79	—	—
1/2	0.36	54.34	—	—
3/4	0.49	90.09	—	—
1	0.61	118.41	—	—
1-1/4	0.69	147.18	—	—
1-1/2	0.77	201.87	—	—
2	0.94	368.37	—	—

Cast Iron #1000 Valves
Threaded and Welded Connections

Diameter	Threaded		Welded	
	Labor	Price	Labor	Price
Ball Valve, 3 pc Full Port				
1/4	0.36	$40.47	—	—
3/8	0.36	40.47	—	—
1/2	0.36	47.15	0.71	$45.33
3/4	0.49	58.54	0.85	57.26
1	0.61	70.92	1.08	71.73
1-1/4	0.69	99.99	1.25	99.99
1-1/2	0.77	134.17	1.51	134.82
2	0.94	180.05	1.93	184.25

Strainer Specialty Valves
Threaded and Flanged Connections

Diameter	Threaded		Flanged	
	Labor	Price	Labor	Price
CS 125 Simplex Basket Strainer				
3/8	0.82	$232.00	—	—
1/2	0.82	232.00	—	—
3/4	0.98	232.00	—	—
1	1.11	313.00	1.11	$454.00
1-1/4	1.23	475.00	1.23	759.00
1-1/2	1.34	475.00	1.34	770.00
2	1.60	650.00	1.41	956.00
2-1/2	2.44	919.00	1.88	1434.00
3	2.83	1203.00	1.92	1535.00
4	—	—	3.06	2175.00
6	—	—	5.23	3988.00
8	—	—	5.62	6438.00
CI 125 Simplex Basket Strainer				
3/8	1.41	$146.00	—	—
1/2	1.41	151.00	—	—
3/4	1.41	192.00	—	—
1	1.57	197.00	1.11	$253.00
1-1/4	1.89	257.00	1.23	316.00
1-1/2	2.05	282.00	1.34	379.00
2	2.29	337.00	1.41	454.00
2-1/2	3.52	451.00	1.88	613.00
3	3.98	547.00	1.92	641.00
4	—	—	3.06	982.00
5	—	—	4.36	1492.00
6	—	—	5.23	1905.00
8	—	—	5.62	3446.00

Strainer Specialty Valves
Threaded and Flanged Connections

Diameter	Threaded		Flanged	
	Labor	Price	Labor	Price
CS Duplex Strainer				
3/4	1.41	$1,747	1.64	$3,681
1	1.57	1,747	1.64	3,681
1-1/4	1.89	2,561	—	—
1-1/2	2.05	2,561	1.89	4,148
2	2.29	4,357	2.25	4,707
2-1/2	—	—	3.46	5,111
3	—	—	3.55	6,826
4	—	—	5.52	9,995
6	—	—	8.33	17,156
8	—	—	8.60	37,151
10	—	—	9.93	45,314
12	—	—	10.56	54,180
14	—	—	11.82	61,916
CI Duplex Strainer				
3/4	1.41	$666	—	—
1	1.57	666	1.64	$1,256
1-1/4	1.89	1,174	1.81	1,306
1-1/2	2.05	1,174	1.89	1,424
2	2.29	1,980	2.25	2,077
2-1/2	3.52	1,980	3.46	2,130
3	—	—	3.55	2,763
4	—	—	5.52	4,350
5	—	—	7.34	8,739
6	—	—	8.33	8,739
8	—	—	8.60	18,613
10	—	—	9.93	25,888
12	—	—	10.56	28,732
14	—	—	11.82	34,808
16	—	—	13.42	—

Strainer Specialty Valves
Threaded and Grooved Connections

Diameter	Threaded		Grooved	
	Labor	Price	Labor	Price
Ductile Iron 250 Wye-Type Strainer				
1/4	0.71	$48.75	—	—
3/8	0.71	48.75	—	—
1/2	0.71	48.75	—	—
3/4	0.90	65.25	—	—
1	0.96	68.75	—	—
1-1/4	1.03	98.50	—	—
1-1/2	1.13	113.25	—	—
2	1.51	173.15	0.61	$149
2-1/2	2.02	556.00	0.60	169
3	2.41	587.25	0.73	197
4	—	—	0.98	360
5	—	—	1.44	562
6	—	—	1.77	686
8	—	—	2.46	1,152
10	—	—	3.32	2,206
12	—	—	4.02	3,310

Strainer Specialty Valves
Flanged and Welded Connections

Diameter	Threaded		Welded	
	Labor	Price	Labor	Price
Steel 300 Wye-Type Strainer				
1/2	0.71	$659	0.57	$635
3/4	0.90	792,	0.84	763
1	0.96	792	1.11	772
1-1/4	1.03	868	1.39	835
1-1/2	1.22	879	1.65	842
2	1.60	742	2.20	742
2-1/2	2.00	883	2.58	883
3	2.26	907	2.72	907
3-1/2	2.44	1,321	3.11	1,321
4	2.62	1,236	3.38	1,236
5	3.19	2,208	3.94	2,208
6	4.38	2,576	4.40	2,576
8	5.42	4,662	5.68	4,662
10	7.15	7,975	7.15	3,536
12	9.10	13,619	9.10	5,264
14	10.46	—	10.46	7,579
16	10.69	—	10.69	—

Gas Specialty Valves
Threaded Connections

Diameter	Threaded	
	Labor	Price
Gas PRV #125 Valve		
3/4	1.36	—
1	1.53	—
1-1/4	1.72	—
1-1/2	1.91	—
2	2.18	—
Gas Cock Valve		
1/2	0.60	—
3/4	0.78	—
1	0.85	—
1-1/4	0.94	—
1-1/2	1.02	—
2	1.18	—
Gas Ball Valve		
3/8	0.60	—
1/2	0.60	—
3/4	0.78	—
1	0.85	—

Steam Trap Specialty Valves

Diameter	Labor	Price
Bucket Trap, Cast Steel		
1/2	0.77	$668
3/4	1.02	1,452
1	1.20	—
1-1/4	1.47	—
1-1/2	1.71	—
2	2.13	—
F & T Trap		
1/2	0.71	$111
3/4	0.94	111
1	1.11	169
1-1/4	1.22	206
1-1/2	1.41	299
2	1.78	545
"Thermostat Trap		
1/4	0.54	—
3/8	0.54	—
1/2	0.54	—
3/4	0.79	—
1	0.93	—
Balanced Pressure Bellows Trap		
1/2	0.78	$185
3/4	0.87	212
Control Disc Trap		
3/8 0.54	$215	—
1/2 0.54	215	—
3/4 0.79	279	—
1 0.93	355	—

Control Specialty Valves

Diameter	Threaded		Flanged	
	Labor	Price	Labor	Price
Set Flow Control Valve				
1/2	0.38	—	—	—
3/4	0.54	—	—	—
1	0.64	—	—	—
1-1/4	0.69	—	—	—
1-1/2	0.71	—	—	—
2	1.01	—	—	—
2-1/2	1.31	—	1.25	—
3	—	—	1.35	—
4	—	—	2.30	—
5	—	—	2.69	—
6	—	—	2.72	—
8	—	—	3.99	—
10	—	—	5.46	—
12	—	—	5.89	—
14	—	—	7.06	—
16	—	—	9.37	—
18	—	—	10.34	—
20	—	—	12.33	—
24	—	—	14.15	—

Diameter	Threaded		Flared	
	Labor	Price	Labor	Price
2-Port Temperature Control Valve with Operator				
1/2	0.85	—	0.74	—
3/4	1.03	—	0.87	—
1	1.19	—	—	—
1-1/4	1.34	—	—	—
1-1/2	1.51	—	—	—
2	1.72	—	—	—

Control Specialty Valves
Threaded and Flanged Connections

Diameter	Threaded		Flanged	
	Labor	Price	Labor	Price
3-Port Temperature Control Valve with Operator				
1/2	1.09	—	—	—
3/4	1.34	—	—	—
1	1.55	—	—	—
1-1/4	1.78	—	—	—
1-1/2	1.93	—	—	—
2	2.30	—	—	—
2-1/2	—	—	3.00	—
3	—	—	3.67	—
4	—	—	6.43	—
5	—	—	7.06	—
6	—	—	7.20	—
8	—	—	7.78	—
10	—	—	13.89	—
12	—	—	14.55	—
14	—	—	16.14	—
16	—	—	17.58	—
Three-Way Control Valve				
1/2	0.37	—	—	—
3/4	0.50	—	—	—
1	0.62	—	—	—
1-1/4	0.73	—	—	—
1-1/2	0.81	—	—	—
2	1.01	—	—	—
2-1/2	1.61	—	—	—
3	1.95	—	—	—

Control Specialty Valves
Flanged Connections

Diameter	Flanged Labor	Flanged Price
Single Temperature Control Butterfly Valve with Operator		
2-1/2	—	—
3	—	—
4	—	—
5	—	—
6	—	—
8	—	—
10	3.00	—
12	3.67	—
14	6.43	—
16	7.06	—
18	7.20	—
20	7.78	—
24	13.89	—
Double Temperature Control Butterfly Valve with Operator		
2-1/2	4.75	—
3	5.78	—
4	7.36	—
5	8.38	—
6	8.81	—
8	11.23	—
10	14.00	—
12	15.04	—
14	18.54	—
16	21.15	—
18	24.04	—
20	29.02	—
24	32.16	—

Pressure and Temperature Specialty Valves
Threaded and Flanged Connections

Diameter	Threaded Labor	Threaded Price	Flanged Labor	Flanged Price
Steam Pressure Regulator Valve 250#				
1/2 1.32	—	—	—	—
3/4 1.57	—	—	—	—
1 1.83	—	—	—	—
1-1/4	2.13	—	—	—
1-1/2	2.39	—	—	—
2 2.79	—	—	—	—
2-1/2	—	—	4.09	—
3 —	—	4.55	—	—
3-1/2	—	—	4.93	—
4 —	—	6.22	—	—
5 —	—	6.94	—	—
6 —	—	8.60	—	—
8 —	—	11.07	—	—
10 —	—	15.10	—	—
12	—	16.52	—	—
Water Pressure Regulator Valve 300#				
1/2 1.00	—	—	—	—
3/4 1.21	—	—	—	—
1 1.42	—	—	—	—
1-1/4	1.59	—	—	—
1-1/2	1.85	—	—	—
2 2.22	—	2.10	—	—
2-1/2	—	—	2.58	—
3 —	—	3.06	—	—
3-1/2	—	—	4.45	—
4 —	—	5.31	—	—
5 —	—	6.48	—	—
6 —	—	7.16	—	—

Pressure and Temperature Specialty Valves
Threaded and Flanged Connections

Diameter	Threaded		Flanged	
	Labor	Price	Labor	Price
Air Pressure Regulator Valve 250#				
1/2	1.08	—	—	—
3/4	1.30	—	—	—
1	1.61	—	—	—
1-1/4	1.95	—	—	—
1-1/2	2.27	—	—	—
2	2.91	—	—	—
2-1/2	—	—	4.15	—
3	—	—	4.61	—
3-1/2	—	—	4.98	—
4	—	—	6.27	—
5	—	—	6.95	—
6	—	—	8.61	—
8	—	—	11.09	—
10	—	—	15.10	—
12	—	—	16.53	—
CI 250# Relief Valve				
1-1/2	—	—	1.71	—
2	—	—	2.06	—
2-1/2	—	—	2.89	—
3	—	—	3.65	—
4	—	—	4.83	—
6	—	—	5.64	—
Bronze Safety Relief Valve				
1/2	0.47	—	—	—
3/4	0.64	—	—	—
1	0.69	—	—	—
1-1/4	0.80	—	—	—
1-1/2	0.97	—	—	—
2	1.14	—	—	—
2-1/2	1.42	—	—	—
3	1.85	—	—	—

Pressure and Temperature Specialty Valves
Threaded and Flanged Connections

Diameter	Threaded		Flanged	
	Labor	Price	Labor	Price
Steam 250# Temperature Regulator Valve				
1/2	1.36	—	—	—
3/4	1.63	—	—	—
1	1.86	—	—	—
1-1/4	2.16	—	—	—
1-1/2	2.41	—	—	—
2	2.94	—	—	—
2-1/2	—	—	4.12	—
3	—	—	4.61	—
3-1/2	—	—	4.98	—
4	—	—	6.27	—
5	—	—	6.95	—
6	—	—	8.61	—
8	—	—	11.09	—
10	—	—	15.10	—
12	—	—	16.53	—
Steam 250# Pressure & Temperature Regulator Valve				
1/2	1.45	—	—	—
3/4	1.70	—	—	—
1	1.95	—	—	—
1-1/4	2.24	—	—	—
1-1/2	2.49	—	—	—
2	3.02	—	—	—
2-1/2	—	—	4.19	—
3	—	—	4.70	—
3-1/2	—	—	5.06	—
4	—	—	6.30	—
5	—	—	6.99	—
6	—	—	8.64	—
8	—	—	11.12	—
10	—	—	15.10	—
12	—	—	16.53	—

Balancing Specialty Valves
Threaded and Flanged Connections

Diameter	Threaded		Flanged	
	Labor	Price	Labor	Price
Circuit Setter				
1/2	0.37	$55	—	—
3/4	0.50	55	—	—
1	0.62	85	—	—
1-1/4	0.73	120	—	—
1-1/2	0.81	140	—	—
2	1.01	200	—	—
2-1/2	1.61	375	—	—
3	1.95	595	—	—
4	—	—	2.27	$835
Balancing Cock Valve				
1/2 0.32	—	—	—	—
3/4 0.45	—	—	—	—
1 0.55	—	—	—	—
1-1/4	0.64	—	—	—
1-1/2	0.73	—	—	—
2 0.94	—	1.49	—	—
2-1/2	1.49	—	1.81	—
3 1.85	—	2.16	—	—
4 —	—	2.67	—	—
5 —	—	3.49	—	—
6 —	—	4.35	—	—
8 —	—	5.30	—	—
10 —	—	7.42	—	—
12 —	—	8.55	—	—
14 —	—	9.56	—	—
16 —	—	9.60	—	—
18 —	—	11.78	—	—
20 —	—	12.59	—	—

Balancing Specialty Valves
Flanged Connections

Diameter	Flanged	
	Labor	Price
Pattern		
2	2.64	$400
2-1/2	3.18	465
3	3.81	505
4	4.72	1,070
5	6.17	1,265
6	7.69	1,420
8	9.34	2,255
10	13.09	3,045
Triple Duty Valve Straight Pattern		
2	2.64	$400
2-1/2	3.18	465
3	3.81	505
4	4.72	1,070
5	6.17	1,265
6	7.69	1,735
8	9.34	2,355
10	13.09	3,885
12	15.09	5,970
14	16.87	7,270
CI 125 Venturi Valve		
4	2.27	—
6	2.94	—
8	4.07	—
10	5.12	—
12	6.78	—
14	7.37	—
16	9.25	—
18	10.47	—
20	11.67	—
24	12.36	—

Balancing Specialty Valves
Grooved Connections

Diameter	Grooved	
	Labor	Price
Venturi Valve		
2-1/2	0.59	—
3	0.71.	—
4	0.94	—
5	1.38	—
6	1.67	—
8	2.26	—
10	2.88	—
12	3.25	—
14	3.50	—
16	3.90	—
18	5.21	—
20	5.71	—
24	7.36	—

Hot Water Specialties
Threaded and Flanged Connections

Diameter	Threaded		Flanged	
	Labor	Price	Labor	Price
Air Separator with Strainer				
2	1.60	$735	—	—
2-1/2	2.44	830	—	—
3	2.83	1,270	—	—
4	—	—	3.06	$1,830
5	—	—	4.36	2,335
6	—	—	5.23	2,795
8	—	—	5.62	4,170
10	—	—	7.57	6,560
12	—	—	8.88	9,430
14	—	—	9.80	19,020
16	—	—	10.72	23,790

Diameter	Threaded		Flanged	
Air Separator without Strainer				
2	1.60	$585	—	—
2-1/2	2.44	680	—	—
3	2.83	960	—	—
4	—	—	3.06	$1,505
5	—	—	4.36	2,005
6	—	—	5.23	2,345
8	—	—	5.62	3,210
10	—	—	7.57	4,900
12	—	—	8.88	7,745
14	—	—	9.80	14,150
16	—	—	10.72	16,030

Hot Water Specialties

Diameter	Labor	Price
Air Purger		
1/2	0.52	—
3/4	0.71	—
1	0.84	—
1-1/4	0.96	—
1-1/2	1.02	—
2	1.15	—
Aerator		
3	2.87	—
4	4.70	—
5	6.94	—
6	12.89	—
Air Vent		
1/2	0.52	$89.00
3/4	0.71	89.00
1	0.84	—
1-1/2	1.02	—
2	1.15	—
Air Bleeder		
1/2	0.52	—
3/4	0.71	—
1	0.84	—
1-1/4	0.96	—
1-1/2	1.02	—
2	1.15	—
2-1/2	1.26	—
3	1.40	—
4	2.37	—
5	3.04	—
6	3.53	—
8	4.76	—
10	6.52	—

Hot Water Specialties
Flanged and Grooved Connections

Diameter	Flanged		Grooved	
	Labor	Price	Labor	Price
Suction Diffuser, 200 PSI				
2	1.41	$279	—	—
2-1/2	1.88	366-	—	—
3	1.92	432	—	—
4	3.06	612	—	—
5	4.36	691	—	—
6	5.23	945	—	—
8	5.62	1,879	—	—
10	7.57	2,639	—	—
12	8.88	3,785	—	—
14	9.80	—	—	—
Suction Diffuser, 300 PSI				
2	1.41	$453	0.89	$268
2-1/2	1.88	543	0.89	359
3	1.92	727	0.89	416
4	3.06	962	1.37	590
6	5.23	1,130	2.77	962
8	5.62	2,233	3.11	1,455
10	7.57	4,206	4.00	2,759

Hot Water Specialties

Diameter	Labor	Price
Cock, Single Threaded		
1/8	0.35	$4.15
1/4	0.35	4.50
3/8	0.35	4.95
1/2	0.35	5.05
Cock, Double Threaded		
1/8	0.35	$5.55
1/4	0.35	5.40
3/8	0.35	5.90
1/2	0.35	7.10
Manual Air Vent		
1/2	0.52	—
3/4	0.71	—
1	0.84	—
1-1/2	1.02	—
2	1.15	—

Refrigeration Specialties

Diameter	Labor	Price
Indicator Sight Glass, Soldered		
1/2	0.59	—
3/4	0.69	—
1	0.76	—
1-1/4	0.81	—
1-1/2	0.84	—
2	0.92	—
Suction Line Filter Drier I Core		
3/4	1.38	—
1	1.53	—
1-1/4	1.67	—
1-1/2	1.74	—
Suction Line Filter Drier 3 Core		
1-1/4	1.72	—
1-1/2	2.18	—
2-1/2	2.31	—
Suction Line Filter Drier 4 Core		
1-1/4	1.78	—
1-1/2	1.86	—
2	2.41	—
3	3.54	—
4	4.30	—
Sealed Filter Drier, Flared		
1/2	0.54	—
3/4	0.54	—
1	0.57	—
Sealed Filter Drier, Soldered		
1/2	1.04	—
3/4	1.21	—
1	1.36	—

Refrigeration Specialties
Refrigeration Valves

	Soldered		Threaded	
Diameter	Labor	Price	Labor	Price
Expansion Valve				
1/2	1.24	—	1.42	—
3/4	1.47	—	1.71	—
1	1.67	—	1.93	—
1-1/4	1.86	—	—	—
1-1/2	1.97	—	—	—
Solenoid Valve Pilot Operated				
1/2	1.24	—	1.36	—
3/4	1.47	—	1.66	—
1	1.65	—	1.88	—
1-1/4	1.78	—	2.12	—
1-1/2	2.31	—	2.74	—
2	2.62	—	3.23	—
2-1/2	3.67	—	4.57	—
Stop Valve				
1/4	—	$12.73	—	—
3/8	—	10.97	—	—
1/2	—	10.97	—	—
3/4	—	13.65	—	—
1	—	19.86	—	—
1-1/4	—	26.95	—	—
1-1/2	—	41.72	—	—
2	—	59.21	—	—
Relief Valve, Soldered				
1/2 0.25	—	—	—	—
3/4 0.32	$36	—	—	—
1 0.42	60	—	—	—
1-1/4	0.47	90	—	—
1-1/2	—	130	—	—
2 —	185	—	—	—

Refrigeration Specialties
Refrigeration ACR Valves

Diameter	Labor	Price
Sight Glass		
5/8	0.59	—
7/8	0.69	—
1-1/8	0.76	—
1-3/8	0.81	—
1-5/8	0.84	—
2-1/8	0.92	—
Filter Dryer		
5/8	1.11	—
7/8	1.32	—
1-1/8	1.53	—
1-3/8	1.62	—
1-5/8	1.69	—
2-1/8	—	—
2-5/8	—	—
3-1/8	—	—
4-1/8	—	—

Steam Specialties

Diameter	Labor	Price
Vacuum Breaker Water-Anti Siphon		
1/2	0.60	$14.85
3/4	0.78	17.55
1	0.85	27.45
1-1/4	0.94	45.50
1-1/2	1.02	53.50
2	1.18	83.20
2-1/2	1.84	240.00
3	2.12	318.00
Cock		
1/8	—	$9.80
1/4	—	9.80
3/8	—	9.70
1/2	—	9.20
3/4	—	10.00
1	—	18.65
1-1/4	—	21.25
1-1/2	—	30.50
2	—	48.00

Steam Specialties

Diameter	Labor	Price
Steam Separator		
1/4	—	$84.00
1/2	—	90.00
3/4	—	95.00
1	—	142.00
1-1/4	—	163.00
1-1/2	—	184.00
2	—	205.00
2-1/2	—	720.00
3	—	815.00
4	—	920.00
5	—	1130.00
6	—	1370.00
8	—	1680.00
10	—	2360.00
12	—	3960.00
14	—	5280.00

Chilled Water Specialties

Diameter	Labor	Price
Sight Glass		
1/2	0.59	—
3/4	0.69	—
1	0.76	—
1-1/4	0.81	—
1-1/2	0.84	—
2	0.92	—

Chapter 19

DWV Pipe and Fittings

This chapter covers labor and price tables for DW (drain, waste and vent) piping and fittings as follows:

DWV Pipe and Fittings:
- Copper DVVV Pipe and Wrot Copper DWV Fittings
- PVC DWV Schedule 40
- ABS DWV Schedule 40
- Cast Iron Hub and Spigot
- Cast Iron No Hub

Units

All diameters are in inches.
Refer to chapter 17 for hangers.

Labor

The labor productivity rates in the tables are what the average mechanic under average conditions can erect on lower floors of construction projects. However, they may require the application of labor correction factors for variable conditions such as upper floors, higher piping runs, congested spaces, etc.

The labor in the tables are in hours per piece and hours per foot for pipe.

Pricing and Discounts

The pricing in the tables are based on the typical manufacturers list prices and the contractor's supplier discounts must be applied.

Appropriate price discounts can range from roughly 10 percent to 80 percent dependent on the following:
- Supplier discount structure
- Who manufacturer is
- Size of batch involved
- Availability
- Shipping distance, costs

Not All List Prices Are Publicly Available

Where there are no prices listed in tables, they are not usually publicly available from the manufacturer. If there is neither labor nor prices listed for a particular diameter in the tables, the particular item may not be available or it may not be a commonly used item.

Caution With List Prices

Pricing can sometimes vary significantly because of changes in commodity material pricing such as copper, etc. Periodic checks of current fist prices should be made.

If in doubt about your discounts from list prices, *check with your supplier* as to what they are for the various categories of items - or request a *quotation*.

Automatic Updating Available in
Wendes Computer System

Automatic updating of list pricing is available in the *Wendes Computerized Piping Estimating System*. Call (847) 808-8371 for more information.

Copper DWV Pipe and Fittings
Soldered Connections

Dia.	Pipe Labor	Pipe Price	1/4 Bend Labor	1/4 Bend Price	1/8 Bend Labor	1/8 Bend Price	1/16 Bend Labor	1/16 Bend Price
1-1/4	0.06	$2.34	0.50	$3.53	0.50	$3.23	0.50	$5.80
1-1/2	0.06	2.95	0.52	4.73	0.52	2.69	0.52	7.62
2	0.06	3.90	0.60	6.87	0.60	6.16	0.60	9.87
3	0.08	6.16	1.15	16.98	1.15	13.04	1.15	16.69
4	0.10	11.10	1.53	80.13	1.53	55.12	1.53	68.00
5	0.15	34.38	—	—	—	—	—	—
6	0.18	47.55	—	—	—	—	—	—
8	0.22	86.42	—	—	—	—	—	—

Dia.	Double Labor	Double Price	Sanitary Tee Labor	Sanitary Tee Price	Dbl Sanitary Tee Labor	Dbl Sanitary Tee Price	Wye Labor	Wye Price
1-1/4	0.71	$14.45	0.71	$6.94	1.03	$17.73	0.71	$13.24
1-1/2	0.76	21.39	0.76	8.64	1.09	15.42	0.76	14.41
2	0.88	33.97	0.88	10.08	1.29	37.42	0.88	19.79
3	—	—	1.70	35.96	2.50	54.50	1.70	45.28
4	—	2.29	91.24	3.39	253.00	2.29	91.24	—
5	—	—	—	—	—	—	—	—
6	—	—	—	—	—	—	—	—

Copper DWV Pipe and Fittings
Soldered Connections

Dia.	Tee-Wye LT Labor	Price	Dbl Wye Labor	Price	P-Trap Labor	Price	Test Cap Labor	Price
1-1/4	0.71	$19.49	—	—	0.50	$23.55	0.25	$0.62
1-1/2	0.76	16.08	1.09	$23.03	0.52	23.47	0.29	0.63
2	0.88	35.65	1.29	37.82	0.60	36.27	0.31	1.36
3	1.70	62.99	2.50	667.16	1.15	87.54	0.59	2.77
4	2.29	165.00	—	—	—	0.77	5.62	—
5	—	—	—	—	—	—	—	—
6	—	—	—	—	—	—	—	—
8	—	—	—	—	—	—	—	—

Dia.	Fitting Plug Labor	Price	Ftg Cleanout Labor	Price	Coupling Labor	Price	Repair Cplg Labor	Price
1-1/4	—	—	0.25	$11.23	0.50	$1.64	0.50	$2.38
1-1/2	—	—	0.29	15.26	0.52	2.05	0.52	2.43
2	0.03	$4.92	0.31	18.06	0.59	2.83	0.59	3.31
3	—	—	0.59	57.43	1.13	5.48	1.13	5.74
4	—	—	0.77	72.25	1.51	13.17	1.51	15.24
5	—	—	—	—	—	—	—	—
6	—	—	—	—	—	—	—	—
8	—	—	—	—	—	—	—	—

PVC DWV Schedule 40
Solvent and Fusion Connections

Dia.	Pipe Labor Solvent	Fusion	Price	1/4 Bend Labor Solvent	Fusion	Price	1/4 Double Bend Labor Solvent	Fusion	Price
1-1/4	0.04	0.04	$1.15	0.16	0.48	$1.77	—	—	—
1-1/2	0.04	0.04	1.26	0.16	0.48	0.89	0.23	0.71	$3.65
2	0.04	0.04	1.64	0.18	0.55	1.38	0.27	0.81	5.11
3	0.05	0.05	3.38	0.24	0.82	3.41	0.34	1.20	15.66
4	0.06	0.06	4.80	0.35	1.07	6.13	—	—	—
6	0.08	0.08	8.89	0.60	1.59	37.29	—	—	—
8	0.10	0.10	13.91	0.82	2.16	75.53	—	—	—
10	—	—	—	1.11	2.94	118.61	—	—	—
12	—	—	—	1.51	3.99	182.45	—	—	—
14	—	—	—	2.06	5.43	244.34	—	—	—
16	—	—	—	2.80	7.39	362.34	—	—	—

Dia.	Return Bend Labor Solvent	Fusion	Price	1/8 Bend Labor Solvent	Fusion	Price	1/6 Bend Labor Solvent	Fusion	Price
1-1/4	—	—	—	0.16	0.48	$1.79	—	—	—
1-1/2	0.16	0.48	$2.95	0.16	0.48	0.99	0.16	0.48	$1.79
2	0.18	0.55	5.67	0.18	0.55	1.44	0.18	0.55	2.68
3	0.24	0.82	15.77	0.23	0.81	2.94	0.23	0.81	9.51
4	0.35	1.07	31.44	0.34	1.07	4.89	0.34	1.07	13.52
6	—	—	—	0.59	1.59	34.13	0.59	1.59	59.91
8	—	—	—	—0.80	2.16	67.86	0.80	2.16	85.48
10	—	—	—	—1.08	2.93	98.72	1.08	2.93	143.24
12	—	—	—	—1.48	3.99	138.65	1.48	3.99	187.67
14	—	—	—	—2.01	5.42	240.43	2.01	5.42	344.00
16	—	—	—	—2.73	7.37	264.22	2.73	7.37	418.23

PVC DWV Schedule 40
Solvent and Fusion Connections

| | 1/16 Bend | | | 1/12 Bend | | | 1/24 & 1/32 Bend | | |
| | Labor | | Price | Labor | | Price | Labor | | Price |
Dia.	Solvent	Fusion		Solvent	Fusion		Solvent	Fusion	
1-1/4	—	—	—	—	—	—	—	—	—
1-1/2	0.16	0.48	$1.47	—	—	—	—	—	—
2	0.18	0.55	1.98	—	—	—	—	—	—
3	0.23	0.81	5.09	—	—	—	—	—	—
4	0.34	1.06	7.74	0.34	1.06	$33.92	0.34	1.06	$19.27
6	0.59	1.59	42.83	0.59	1.59	52.96	0.59	1.59	31.01
8	0.80	2.16	52.43	0.80	2.16	76.54	0.80	2.16	52.43
10	1.08	2.93	79.12	1.08	2.93	115.27	1.08	2.93	79.12
12	1.48	3.99	125.16	1.48	3.99	141.27	1.48	3.99	125.16
14	2.01	5.42	187.58	2.01	5.42	238.64	2.01	5.42	196.77
16	2.73	7.37	255.60	2.73	7.37	329.52	2.73	7.37	268.24

| | Vent Tee | | | Sanitary Tee | | | Wye | | |
| | Labor | | Price | Labor | | Price | Labor | | Price |
Dia.	Solvent	Fusion		Solvent	Fusion		Solvent	Fusion	
1-1/4	—	—	—	0.22	0.71	$3.15	0.23	0.71	$3.02
1-1/2	0.22	0.71	$4.98	0.22	0.71	1.32	0.23	0.71	2.62
2	0.26	0.81	3.51	0.26	0.81	1.81	0.27	0.81	2.63
3	0.34	1.20	13.15	0.34	1.20	4.99	0.34	1.20	5.97
4	0.50	1.59	15.79	0.50	1.59	9.51	0.51	1.60	11.12
6	—	—	—	0.87	2.37	55.31	0.87	2.39	55.66
8	—	—	—	1.20	3.26	165.55	1.23	3.39	119.16
10	1.81	4.94	173.34	1.81	4.94	416.80	1.81	4.98	164.23
12	2.75	7.50	230.36	2.75	7.50	476.40	2.75	7.57	243.27
14	3.44	9.37	452.82	3.44	9.37	834.82	3.44	9.45	477.25
16	4.29	11.69	640.96	4.29	11.69	1161.66	4.29	11.80	548.13

PVC DWV Schedule 40
Solvent and Fusion Connections

Dia.	Double Wye Labor Solvent	Double Wye Labor Fusion	Double Wye Price	Slip Cross Labor Solvent	Slip Cross Labor Fusion	Slip Cross Price	P-Trap w/Chrome Nut Labor Solvent	P-Trap w/Chrome Nut Labor Fusion	P-Trap w/Chrome Nut Price
1-1/4	—	—	—	—	—	—	—	—	—
1-1/2	0.29	0.94	$5.58	—	—	—	0.17	0.48	$2.57
2	0.34	1.06	7.18	—	—	—	0.19	0.56	3.49
3	0.44	1.59	18.53	—	—	—	0.19	0.82	17.40
4	0.66	2.11	37.56	—	—	—	0.19	1.08	42.00
6	1.21	3.15	73.85	1.17	3.15	$70.28	—	—	—
8	1.65	4.30	112.54	1.58	4.26	123.31	—	—	—
10	2.50	6.50	206.33	2.39	6.44	231.07	—	—	—
12	3.79	9.86	336.84	3.63	9.77	304.62	—	—	—
14	4.75	12.34	496.01	4.54	12.20	506.04	—	—	—
16	5.92	15.40	627.33	5.66	15.23	739.22	—	—	—

Dia.	Adapter HxF Labor Solvent	Adapter HxF Labor Fusion	Adapter HxF Price	Adapter HxM Labor Solvent	Adapter HxM Labor Fusion	Adapter HxM Price	Expansion Joint Labor Solvent	Expansion Joint Labor Fusion	Expansion Joint Price
1-1/4	—	—	—	—	—	—	—	—	—
1-1/2	0.29	0.94	$5.58	—	—	—	0.17	0.48	$2.57
2	0.34	1.06	7.18	—	—	—	0.19	0.56	3.49
3	0.44	1.59	18.53	—	—	—	0.19	0.82	17.40
4	0.66	2.11	37.56	—	—	—	0.19	1.08	42.00
6	1.21	3.15	73.85	1.17	3.15	$70.28	—	—	—
8	1.65	4.30	112.54	1.58	4.26	123.31	—	—	—
10	2.50	6.50	206.33	2.39	6.44	231.07	—	—	—
12	3.79	9.86	336.84	3.63	9.77	304.62	—	—	—
14	4.75	12.34	496.01	4.54	12.20	506.04	—	—	—
16	5.92	15.40	627.33	5.66	15.23	739.22	—	—	—

PVC DWV Schedule 40
Solvent and Fusion Connections

Dia.	Slip Cap Labor Solvent	Slip Cap Labor Fusion	Slip Cap Price	Spigot Plug Labor Solvent	Spigot Plug Labor Fusion	Spigot Plug Price	Coupling HxH Labor Solvent	Coupling HxH Labor Fusion	Coupling HxH Price
1-1/4	—	—	—	—	—	—	0.16	0.48	$1.70
1-1/2	0.10	0.25	$2.26	0.10	0.25.	$6.19	0.16	0.48	0.57
2	0.11	0.29	3.75	0.11	0.29	10.17	0.18	0.55	0.59
3	0.13	0.34	6.51	—	—	—	0.23	0.81	1.70
4	0.17	0.43	9.30	—	—	1	0.34	1.06	2.79
6	0.28	0.78	23.48	0.28	0.73	21.23	0.57	1.59	14.94
8	0.42	—	—	0.42	1.09	54.22	0.78	1.59	33.64
10	0.70	1.82	54.92	0.70	1.82	61.83	0.99	2.02	38.61
12	1.06	2.76	69.35	1.06	2.76	79.77	1.20	2.45	43.67
14	1.33	3.46	117.92	1.33	3.46	117.92	1.45	2.97	100.61
16	1.66	4.31	150.96	1.66	4.31	183.24	1.76	3.60	112.21

ABS DWV Schedule 40
Solvent and Fusion Connections

Dia.	Pipe Labor Solvent	Pipe Labor Fusion	Pipe Price	1/4 Bend Labor Solvent	1/4 Bend Labor Fusion	1/4 Bend Price	1/4 Double Bend Labor Solvent	1/4 Double Bend Labor Fusion	1/4 Double Bend Price
1-1/4	—	—	—	0.16	0.48	$2.05	—	—	—
1-1/2	0.06	—	$2.70	0.16	0.48	1.09	0.23	0.71	$4.22
2	0.08	—	3.45	0.18	0.55	1.79	0.27	0.81	5.90
3	0.10	—	6.90	0.24	0.82	4.24	0.34	1.20	18.08
4	0.13	—	9.95	0.35	1.07	7.38	—	—	—
6	0.17	—	19.25	0.60	1.59	46.57	—	—	—

Dia.	Return Bend Labor Solvent	Return Bend Labor Fusion	Return Bend Price	1/8 Bend Labor Solvent	1/8 Bend Labor Fusion	1/8 Bend Price	1/6 Bend Labor Solvent	1/6 Bend Labor Fusion	1/6 Bend Price
1-1/4	—	—	—	0.16	0.48	$2.07	—	—	—
1-1/2	0.16	0.48	$3.35	0.16	0.48	1.17	0.16	0.48	$1.96
2	0.18	0.55	5.83	0.18	0.55	1.74	0.18	0.55	3.14
3	0.24	0.82	16.70	0.23	0.81	3.88	0.24	0.81	11.00
4	0.35	1.07	35.79	0.34	1.07	6.19	0.35	1.07	16.09
6	—	—	—	0.59	1.60	37.21	—	—	—

Dia.	1/6 Bend Labor Solvent	1/6 Bend Labor Fusion	1/6 Bend Price	Vent Tee Labor Solvent	Vent Tee Labor Fusion	Vent Tee Price	Sanitary Tee Labor Solvent	Sanitary Tee Labor Fusion	Sanitary Tee Price
1-1/4	—	—	—	—	—	—	0.22	0.71	$2.58
1-1/2	0.16	0.48	$1.71	0.22	0.48	$4.47	0.22	0.71	1.60
2	0.18	0.55	2.28	0.26	0.55	4.06	0.26	0.81	2.27
3	0.23	0.81	5.88	0.34	0.82	15.19	0.34	1.20	5.76
4	0.34	1.06	8.96	0.50	1.59	19.96	0.50	1.59	14.29
6	0.59	1.60	72.75	—	—	—	0.87	2.37	68.94

ABS DWV Schedule 40
Solvent and Fusion Connections

	Wye			Double Wye			Expansion Joint		
	Labor		Price	Labor		Price	Labor		Price
Dia.	Solvent	Fusion		Solvent	Fusion		Solvent	Fusion	
1-1/4	0.23	0.71	$2.98	—	—	—	—	—	—
1-1/2	0.23	0.71	3.03	0.29	0.94	$6.44	—	—	—
2	0.27	0.81	3.11	0.34	1.06	8.29	0.18	0.55	$193.87
3	0.34	1.20	7.16	0.44	1.59	21.43	0.24	0.82	211.61
4	0.51	1.60	16.97	0.66	2.11	43.41	0.35	1.07	349.70
6	0.87	2.39	64.33	—	—	—	—	—	—

	Adapter HxF			Adapter HxM			Coupling		
	Labor		Price	Labor		Price	Labor		Price
Dia.	Solvent	Fusion		Solvent	Fusion		Solvent	Fusion	
1-1/4	—	—	—	0.33	0.59	$1.44	0.16	0.48	$1.15
1-1/2	0.36	0.64	$1.20	0.36	0.64	1.04	0.16	0.48	0.68
2	0.43	0.76	1.93	0.43	0.76	1.52	0.18	0.55	0.71
3	0.78	1.25	5.25	0.78	1.25	4.43	0.23	0.81	2.38
4	1.03	1.66	6.61	1.03	1.66	9.81	0.34	1.06	3.65
6	1.57	1.66	22.55	—	—	—	0.57	1.59	19.30

	P-Trap			Cap Hub			Pipe Test Cap		
	Labor		Price	Labor		Price	Labor		Price
Dia.	Solvent	Fusion		Solvent	Fusion		Solvent	Fusion	
1-1/4			—	—	—	—	—	—	—
1-1/2	0.17	0.48	$3.31	0.03	0.03	$2.36	0.03	0.03	$0.65
2	0.19	0.56	5.36	0.03	0.03	3.78	0.03	0.03	0.6
3	0.19	0.82	24.51	0.03	0.03	6.70	0.03	0.03	0.86
4	0.19	1.08	50.21	0.03	0.03	8.98	0.03	0.03	1.09
6	—	—	—	—	—	—	—	—	—

Cast Iron Hub and Spigot
O-Ring Joint Compression Connections

Dia.	Pipe 10' Section		Pipe 5' Section		1/4 Bend		1/8 Bend	
	Labor	Price	Labor	Price	Labor	Price	Labor	Price
2	0.98	$52.80	0.59	$44.35	0.29	$9.70	0.29	$7.75
3	1.26	72.80	0.80	49.20	0.43	16.80	0.42	14.10
4	1.61	94.65	1.01	63.95	0.58	26.25	0.56	20.50
5	1.89	133.50	1.22	82.90	0.71	36.65	0.69	28.35
6	2.24	162.65	1.47	120.80	0.79	45.70	0.75	34.80
8	3.08	260.95	1.96	181.55	1.08	137.75	1.02	103.90
10	3.99	434.45	2.59	275.95	1.48	201.40	1.36	149.40
12	4.41	631.05	3.11	364.25	1.78	272.85	1.62	283.20
15	5.04	922.35	3.78	540.80	2.81	874.80	2.37	565.00

Dia.	1/5 Bend		1/6 Bend		1/16/Bend		Pipe Plug	
	Labor	Price	Labor	Price	Labor	Price	Labor	Price
2	0.29	$15.95	0.29	$9.70	0.29	$7.75	0.04	$5.15
3	0.42	24.80	0.42	16.70	0.42	12.90	0.06	9.05
4	0.57	36.85	0.57	23.15	0.55	18.00	0.06	10.95
5	0.71	47.90	0.71	38.35	0.66	27.05	0.08	21.20
6	0.78	52.05	0.78	54.10	0.73	30.30	0.10	20.60
8	—	—	1.07	124.75	0.99	106.40	0.15	50.50
10	—	—	1.29	247.70	1.29	163.55	0.21	71.10
12	—	—	—	—	1.62	384.75	0.31	103.80
15	—	—	—	—	2.37	449.10	0.41	178.20

Pipe pricing and labor for Cast Iron Hub and Spigot is per section. Divide section length to get per foot price and labor values.

Cast Iron Hub and Spigot
O-Ring Joint Compression Connections

Dia.	Wye Labor	Wye Price	Dbl Wye Labor	Dbl Wye Price	Sanitary Tee Labor	Sanitary Tee Price	Sanitary Cross Labor	Sanitary Cross Price
2	0.54	$16.05	0.77	$26.55	0.54	$15.45	0.77	$24.45
3	0.82	29.55	1.17	45.95	0.81	28.35	1.15	42.45
4	1.07	39.70	1.57	63.70	1.06	334.80	1.53	58.00
5	1.34	66.90	1.95	102.80	1.30	69.15	1.80	83.60
6	1.49	91.40	2.16	163.80	1.46	78.50	2.02	136.10
8	1.99	223.50	2.87	401.05	1.93	228.10	2.67	261.10
10	2.75	361.25	3.88	563.40	2.58	381.50	—	—
12	3.28	621.05	4.59	812.70	3.04	635.35	—	—
15	4.15	1316.10	5.10	2449.50	4.15	1429.95	—	—

Dia.	Spigot Vent Cap Labor	Spigot Vent Cap Price	Hub End Vnt Cp Labor	Hub End Vnt Cp Price	Vent Cap w/Screw Labor	Vent Cap w/Screw Price	Dbl Hub Ftg Labor	Dbl Hub Ftg Price
2	—	—	—	—	0.04	$7.70	0.50	$11.65
3	—	—	—	—	0.06	10.90	0.73	19.30
4	0.06	$28.60	0.06	$29.70	.0.06	19.50	0.98	21.95
5	—	—	—	—	0.08	25.30	1.20	42.15
6	0.10	47.85	—	—	—	—	1.30	43.20
8	—	—	—	—	—	—	1.66	107.15
10	—	—	—	—	—	—	2.25	238.80
12	—	—	—	—	—	—	2.56	340.55
15	—	—	—	—	—	—	3.16	475.95

Cast Iron Hub and Spigot
O-Ring Joint Compression Connections

Dia.	Clean Out Labor	Clean Out Price	P-Trap Labor	P-Trap Price	Running Trap Labor	Running Trap Price
2	0.04	$5.15	0.33	$20.00	0.33	$25.35
3	0.06	8.10	0.48	29.55	0.50	42.95
4	0.06	15.55	0.64	43.15	0.66	79.75
5	0.08	22.10	0.94	91.05	0.85	250.10
6	0.10	28.15	1.42	132.95	0.97	368.75
8	0.15	38.95	2.10	378.90	—	—
10	0.21	66.95	—	—	—	—
12	—	—	—	—	—	—
15	—	—	—	—	—	—

Cast Iron No Hub
No Hub Connections

Dia.	Pipe		1/4 Bend		Dbl 1/4 Bend		1/8 Bend	
	Labor	Price	Labor	Price	Labor	Price	Labor	Price
1-1/2	0.08	$5.16	0.21	$7.10	—	—	0.21	$5.90
2	0.08	5.28	0.27	7.70	0.39	$17.25	0.27	6.55
3	0.10	7.28	0.40	10.70	0.58	26.40	0.38	8.85
4	0.14	9.47	0.52	15.40	0.78	34.70	0.50	11.25
5	0.16	13.35	0.67	34.45	—	—	0.63	24.05
6	0.19	16.27	0.81	38.80	—	—	0.76	26.10
8	0.27	26.10	0.99	106.95	—	—	0.88	75.20
10	.36	43.45	—	—	—	—	1.23	140.05

Dia.	1/5 Bend		1/6 Bend		1/16 Bend		Blind Plug	
	Labor	Price	Labor	Price	Labor	Price	Labor	Price
1-1/2	—	—	—	—	0.20	$6.15	0.12	$3.70
2	0.27	$9.75	0.27	$8.00	0.27	6.75	0.15	3.70
3	0.39	16.95	0.39	9.80	0.38	8.00	0.20	5.55
4	0.52	26.10	0.52	13.55	0.50	9.25	0.27	8.55
5	—	—	—	—	0.62	22.85	0.34	16.00
6	—	—	—	—	0.75	27.70	0.40	16.65
8	—	—	—	—	0.85	43.95	0.44	44.50
10	—	—	—	—	—	—	0.49	65.50

Cast Iron No Hub
No Hub Connections

Dia.	Sanitary Tee Labor	Sanitary Tee Price	Sanitary Cross Labor	Sanitary Cross Price	Wye Labor	Wye Price	Double Wye Labor	Double Wye Price
1-1/2	0.29	$9.80	0.39	$14.10	0.29	$10.00	—	—
2	0.39	10.70	0.50	17.85	0.38	10.00	0.50	$15.40
3	0.57	13.00	0.75	27.65	0.57	14.20	0.75	28.35
4	0.76	20.15	1.03	52.30	0.77	23.10	1.03	57.90
5	1.00	57.35	—	—	0.99	53.60	—	—
6	1.15	58.50	1.27	110.55	1.19	61.60	1.57	102.10
8	1.58	237.05	1.94	241.10	1.47	145.00	1.94	291.50
10	—	—	—	—	2.10	315.35	—	—

Dia	P-Trap Labor	P-Trap Price	Coupling w/Band Labor	Coupling w/Band Price
1-1/2	0.22	$11.75	0.20	$6.95
2	0.29	11.25	0.27	6.95
3	0.42	24.90	0.38	8.25
4	0.60	43.15	0.50	9.85
5	—	—	0.62	24.05
6	1.03	104.60	0.75	25.05
8	—	—	0.85	47.25
10	—	—	1.23	65.05

This page intentionally left blank

Section VI

Contracting for Profit

This page intentionally left blank

Chapter 20

Markups for Overhead and Profit

UNDERSTANDING AND APPLYING CORRECT OVERHEAD AND PROFIT FACTORS

Most of the manual to this point has been spent on emphasizing the necessity for making sure direct material and labor costs are covered fully and accurately. However, you may cover direct costs correctly, and still may be short many thousands of dollars on your bid by putting an inadequate markup on the job for overhead and profit thus despite your success in the first part, you are really no where in the end.

Hence, not only must you make sure that the estimate of labor and materials is complete and accurate-but also that the profit and overhead markups on each job bid correspond accurately to the actual finances of the company.

Hence, the following concepts and definitions must be thoroughly understood in the preparation of bids:

DIRECT COSTS

Direct costs are the costs for material, labor, and sub-contractors needed for a job. They are charged directly to the particular project.

Labor covers fabrication, installation, drafting, balancing, etc.; materials include raw material and equipment; sub-contractors include rentals and these are all normally direct costs. Also, sales tax, permits, bonds, travel costs, room and board are direct type costs which can be attributed directly to a project.

Total direct costs for the year for the $2 million dollar sales contractor in the sample illustration are $1,500,000, which is 75 percent of sales.

OVERHEAD COSTS

Overhead costs however, are those costs you cannot charge directly to each specific project. They are fixed and semi-variable costs that continue on, either 100 percent of the time or to some degree, whether you have work or not. Overhead costs are prorated in varying degrees to all projects. Some of the main ones are:

Office Salaries	Interest
Rent or Owning Costs	Depreciation
Utilities	Insurance
Real Estate Taxes	Boats, Lear Jets
Auto and Truck Costs	Trips to Disney World

Labor draws about two to three times as much indirect cost as equipment, materials and sub-contractors do.

Total overhead costs for the year in the sample illustration are $400,000, which is 20 percent of sales.

PROFIT

Profit is the amount of money you have left over at the end of the year and at the completion of the projects after you have paid out all your direct and overhead costs. Net profits are after corporate income taxes are taken out.

Honest profit is not a dirty word, but rather a necessity for successful business operation in a free functioning economy:

- Profit pays a return on investment for stockholders
- It provides money for capital investments, machinery, buildings, etc.
- It provides for replacement, improvements and growth.
- It provides for increases in wages and more fringe benefits for employees
- Profit provides more jobs through expansion
- Profit is the incentive needed to take the risks of a business, and is the reward for the hard work and risk.
- Profits are necessary for a healthy economy.

- Profits pay taxes to the government

A business cannot survive long without profits, and the if the business fails, many parties are hurt, not only the owners, but also employees, creditors, customers, etc.

A company is in business to make money not lose it. Its primary goal is to earn a sufficient amount of profit to cover the items listed above.

However, just because someone is in business and must make money to survive, it also doesn't justify, on the other hand, a carte blanche approach. A business must run responsibly in the context of the overall good of the community, etc. and must meet the following criteria:

- The products and services of a company must meet acceptable standards and satisfy needs.
- The business cannot break the law.
- It must meet common morality standards
- It must provide healthy and safe working conditions for its employees.
- It should not adversely effect the environment
- It cannot adversely effect other peoples property
- It cannot adversely effect the health or safety of the public in general.

Profit for the year in the sample illustration is $100,000, which is 5 percent of sales.

RETURN ON INVESTMENT

Given a $2 million sales and a $500,000 capital investment in a company:
- If a 20% return on investment is required, (which is $100,000 based on the $500,000 capital investment), there must be a 5% profit of $ 100,000 on the sales of $2 million.
- A corporate tax of 25% reduces the profit to $75,000.
- If $50,000 is spent on new machinery etc. it reduces the net profit to $25,000 or earnings of 2-1/2% for stockholders.

MARKUP FOR OVERHEAD

Markup for overhead is the amount of money added to direct costs on an estimate to cover overhead costs.

The markup percentage is the ratio to direct costs.

Average percentage markup required on jobs during the year:

$400,000 (yearly overhead costs)/$S1,500,000 (yearly direct costs) 27% average (rounded out).

The goal of an overhead markup on the various projects is to recuperate all the actual overhead costs incurred for the year. Each job must make its proportionate contribution to overhead and plus contribute proportionately to the profit.

MARGIN, GROSS PROFITS

Margin, which is also called gross profit, is the difference between sales and direct costs,and is equal to the overhead and profit together.

The margin percentage is the ratio to total sales, as differentiated from markup, which is the ratio to total direct costs. However, they are both the same thing money wise, but just have different names and different ratio references.

Margin works from the top down from total sales, as an accountant does, rather than from direct costs up as the estimator does. Margin, obviously then, is always a lower percentage than markup since it is a ratio to a higher base. Hence:

The margin in the illustration is $500,000 and includes the overhead and profit for the year.

The $500,000, as a ratio to $2 million sales, is 25%.

The $500,000, as a ratio to direct costs to $1,500,000, is 33 %.

METHODS OF DETERMINING PERCENTAGE MARKUPS ON COSTS

Single Markup On Labor and Materials for Overhead

A single markup for overhead, which is a percentage of the combined total of direct labor and material costs, is the basic and clearest approach in marking up jobs. Whether estimating or monitoring jobs, when you have them, you should always know what the average percentage should be. It is a starting and reference point and gives you a vital overall view.

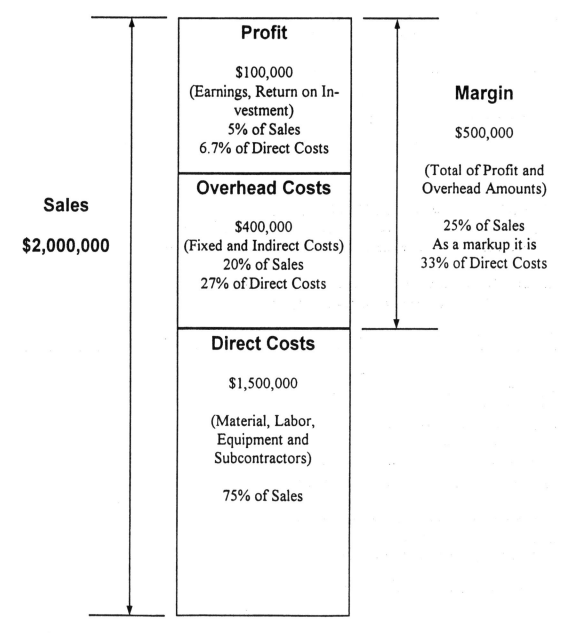

An estimator, when preparing a bid, works up the direct costs on the project first and then "markups" to cover overhead and profit as a ratio to direct costs. The accountant, in his financial statements, deals with the percentage overhead and profit as a ratio to the total sales figure.

$$\frac{\$400,000 \text{ overhead}}{\$1,500,000 \text{ direct costs}} = \begin{array}{c} 27\% \text{ average markup} \\ \text{required on} \\ \text{direct costs} \end{array}$$

$$\frac{\$500,000 \text{ overhead and profit}}{\$1,5000,000} = \begin{array}{c} 33\% \text{ average markup} \\ \text{required on} \\ \text{direct costs} \end{array}$$

PROFIT MARKUPS

The markup for profit should generally be kept separate from the overhead markup so that overhead can be covered more accurately and distinctly in the bid. Then, when it is necessary to trim the price on a bid to be more competitive, it is easier to only focus on reducing the profit, rather than the overhead.

MULTIPLE MARKUPS

The multiple markup method puts a different percentage markups on labor and material. Even though a single markup on material and labor is easy to apply and correlates well to financial statements, it frequently is not a competitive approach to bidding because it may not place the bulk of the overhead on where it properly belongs, which is on labor. As stated previously, since labor generally draws more overhead than equipment and materials, it must contribute a larger portion to it.

Consequently when applying a multiple markup, a lower percentage is applied on equipment, and a higher one on labor. This more properly and competitively covers the variations in material/labor ratios and is more realistic. The approach here is to select an appropriately lower percentage markup on equipment and subs first, according to your market conditions and material/labor ratios of the project being bid. Then put the balance of the overhead burden on labor.

For example:

Given: Labor Costs $750,000
         Equipment and Subs $750,000

A 10% markup on equipment and subs for overhead is $75,000,

$400,000   Total overhead
 − 75,000   Less equipment/subs overhead markup
$325,000   Overhead left for labor to cover

Markup percentage needed on labor

$325,000 (Labor Overhead Burden)
$750,000 (Labor Costs)   *Equals a 50% markup needed on labor for overhead*

MARKUPS ON LABOR HOURS ONLY

Just adding a markup to direct labor costs, to cover overhead and profit, can be a simpler, more convenient approach for change orders, time and material orders and possibly for certain type operations according to the preference of the contractor or customer.

For example:

Given: 9 mechanics employed in company = 17,857 man hours/year

If all overhead is covered by labor:

$$\frac{\$400,000 \text{ Total Overhead}}{17,857 \text{ Hours}} = \$22.40 \text{ which must be added onto direct labor costs}$$

$$\frac{\$100,000 \text{ Total Profit}}{17,857 \text{ Hours}} = \$5.60 \text{ which must be added on also}$$

Direct labor costs with fringes, pay
      roll taxes, etc. ....$42.00 per hr
Overhead costs $22.40
Plus 5% Profit..$ 5.60
Total required selling price per hr for labor $70.00

BIDDING STRATEGY REGARDING MARKUPS

You must know what average percentage markup is needed on direct costs during the course of the year to properly cover the actual overhead and profit final amount at year end. It must be based on anticipated sales, overhead costs and desired profit and be monitored monthly and adjusted if need be.

Even though you may have to vary your percentages at times for various type projects, bidding situations and material labor ratios, the overall average must come out to the percentage determined by yearly sales and overhead.

Overhead must be totally covered for the year no matter what variation in percentages are applied on the estimates.

Each job must make its proper proportionate contribution to both overhead and profit and the total must equal the years total overhead costs and meet the anticipated profit.

Situations where You May Vary Your Standard Markup
- The markup for overhead might be varied where the **number of bidders** is considerably more or less than usual, or where the competitors may change the bidding strategy.

- Jobs which are considerably **larger** or **smaller than your average.**

- Where the **ratio of materials and subcontractors to labor** is greater or less than normal.

- For **low risk or higher risk** projects than the normal, or those with considerable capital involved.

Markup Calculation Sheet

COMPANY MARKUP CALCULATION SHEET

Date *November 1, 1997*

Company *Northwest Mechanical*   Period *Nov. 1997 to Nov. 1998*

		Percent of Sales
1. Anticipated Sales for a Year	$ *2,000,000*	
2. Total Indirect Overhead and Administration Costs for the Year	$ *400,000*	*20%*
Profit Desired _5_ %	$ *100,000*	*5%*
Total Anticipated Direct Costs for the Year (Material and Labor)	$ *1,500,000*	*75%*
Breakdown: Labor (Includes fringes, payroll taxes, insur.)	$ *750,000*	
Material and Equipment	$ *750,000*	
Subs	$	

3. SINGLE MARKUP NEEDED ON TOTAL DIRECT COSTS

Percent For Ovhd Only:	$ Overhead Costs	$ *400,000*			
	$ Direct Costs	$ *1,500,000*	=	*27*	%

Percent	$ Overhead & Profit	$ *500,000*			
For Overhead & Profit Together	$ Direct Costs	$ *1,500,000*	=	*33*	%

4. SIMPLIFIED DUAL MARKUP FOR OVERHEAD

Amount of Markup For Year

Markup on Materials and Equip	_10_ % x $ *750,000*	=	$ *75,000*
Markup on Subs	___ % x $_____	=	$
Total Overhead on Mat. & Subs			$ *75,000*

Percent
Markup on Labor = ($ Total Ovhd) - ($Matl. & Sub Ovhd)
for Overhead   (Labor Costs)

$$\frac{\$(\ 400,000 \ ) - \$(\ 75,000 \ )}{\$(\ \ 750,000 \ \ )} = \frac{\$ \ 325,000}{\$ \ 750,000} = (\quad 43 \quad)\%$$

5. TOTAL SELLING COST OF LABOR PER HR COVERING OVERHEAD AND PROFIT

Wages per hr (incl. fringes, insr., & taxes)	$ *42.00*	
$ Wages x percent overhead markup on labor	$ *22.40*	*(17.857 hr)*
Profit _5_ % (on total sales)	$ *5.60*	
TOTAL	$ *70.00*	

The above markup calculation sheet can be used for determining the required level of markups needed for overhead and profit for a company using the three methods described in this chapter: 1) a single markup applied to total direct costs either for overhead alone or for the combination of overhead and profit as one markup factor. 2) individual, multiple markups for labor, materials and subcontractors. 3) selling cost of labor.

Single Markup Factors On Direct Costs
To Cover Both Overhead and Desired Profit
As Percent of Sales

Overhead Margin as Percent of Sales	DESIRED NET PROFIT								
	4%	5%	6%	7%	8%	9%	10%	15%	20%
10%	1.16	1.18	1.19	1.20	1.22	1.23	1.25	1.33	1.43
11%	1.18	1.19	1.20	1.22	1.23	1.25	1.27	1.35	1.45
12%	1.19	1.20	1.22	1.23	1.25	1.27	1.28	1.37	1.47
13%	1.20	1.22	1.23	1.25	1.27	1.28	1.30	1.39	1.49
14%	1.22	1.23	1.25	1.27	1.28	1.30	1.32	1.41	1.52
15%	1.23	1.25	1.27	1.28	1.30	1.32	1.33	1.43	1.54
16%	1.25	1.27	1.28	1.30	1.32	1.33	1.35	1.45	1.56
17%	1.27	1.28	1.30	1.32	1.33	1.35	1.37	1.47	1.59
18%	1.28	1.30	1.32	1.33	1.35	1.37	1.39	1.49	1.61
19%	1.30	1.32	1.33	1.35	1.37	1.39	1.41	1.52	1.64
20%	1.32	1.33	1.35	1.37	1.39	1.41	1.43	1.54	1.67
21%	1.33	1.35	1.37	1.39	1.41	1.43	1.45	1.56	1.69
22%	1.35	1.37	1.39	1.41	1.43	1.45	1.47	1.59	1.72
23%	1.37	1.39	1.41	1.43	1.45	1.47	1.49	1.61	1.75
24%	1.39	1.41	1.43	1.45	1.47	1.49	1.52	1.64	1.79
25%	1.41	1.43	1.45	1.47	1.49	1.52	1.54	1.67	1.82
26%	1.43	1.45	1.47	1.49	1.52	1.54	1.56	1.69	1.85
27%	1.45	1.47	1.49	1.52	1.54	1.56	1.59	1.72	1.89
28%	1.47	1.49	1.52	1.54	1.56	1.59	1.61	1.75	1.92
29%	1.49	1.52	1.54	1.56	1.59	1.61	1.64	1.79	1.96
30%	1.52	1.54	1.56	1.59	1.61	1.64	1.67	1.82	2.00

Example: If overhead is *20%* of sales and you are shooting for a net profit of *5%* before taxes, use *1.33* as a single markup on direct costs.

- Where you find you will be **closing more work during the year** than you had anticipated; and where overhead will be greater than anticipated; where **gross margins on projects are greater** than calculated for the year, and you find you have overhead covered ahead of time.

- **Potential extras,** tenant areas, additions, which produce higher gross margins may influence your markup.

- Fast moving, **quick completion projects** may reduce overhead costs somewhat.

- **Quick paying jobs.**

- **Anticipated buyouts** on equipment and sub-contractors may reduce the percentage markup required.

- Large contracts where the cost of money may increase the markup percentage needed.

- The **markup may be adjusted according to the estimators general accuracy.** His estimates may be generally higher or lower than actual direct costs.

ODDS OF GETTING JOB

The odds of being low on a bid and getting the contract are contingent on the following factors:

- Number of bidders.
- Whether being bid as a prime contractor or sub-contractor.
- Bidding to owner, architect/engineer or general contractor.
- The efficiency and productivity of your company compared to competitors.

- Percentage markup you require for overhead and profit compared to competitors.
- How badly you are in need of work compared to competitors.
- The general accuracy of estimating of the other contractors
- How rationally the competitors normally bid.

Income Statement
12 Months Ending December 31, 1997
(Typical $2 Million Mechanical Contractor)

	Amount	Percent of Sales
Sales	$2,000,000	100.00
Direct Costs (variable with volume of work)		
Material	700,000,	35.00
Sub Contracts	100,000	5.00
Labor	500,000	25.00
Fringes, Insurance, Taxes	200,000	
Total Direct Costs	1,500,000	75.00
Gross Margin	500,000	25.00
Overhead Expenses (fixed and semi fixed costs)		
Salaries Officers	120,000	6.00
Office	60,000	3.00
Fringe, Insurance, Taxes	40,000	2.00
Rent, 6000 square feet	30,000	1.50
Utilities	10,000	0.50
Insurance - Fire and Theft	20,000	1.00
Taxes - Real Estate	8,000	0.40
Interest	16,000	0.80
Bad Debts	4,000	0.20
Maintenance	8,000	0.40
Expendable Tools	6,000	0.30
Office Supplies	8,000	0.40
Dues	1,000	0.05
Donations	1,000	0.05
Advertising	2,000	0.10
Professional Expenses	4,000	0.20
Auto and Truck, Owning & Operating	20,000	1.00
Sales Expenses	4,000	0.20
Travel and Entertainment	2,000	0.10
Depreciation - Auto and Trucks	20,000	1.00
Depreciation - Machinery and Equipment	10,000	0.50
Depreciation - Building	6,000	0.30
Total Overhead	$400,000	20.00
Net Income Before Income Taxes	$100,000	5.00

Typical Income Statement

Balance Sheet
December 31,1997

Assets

Current Assets		
Cash	$100,000	
Accounts Receivables	200,000	
Inventory	120,000	
Work in progress	60,000	
Prepaid expenses	10,000	
Total current assets	$580,000	
Fixed Assets		
Buildings	$300,000	
Machinery & Equipment	200,000	
Autos and Trucks	60,000	
Office Equipment	10,000	
Total	$570,000	
Less Depreciation	−100,000	
	$470,000	
Plus Land	30,000	
Total Fixed Assets		$500,000
Other Assets		
Deposits, Securities		20,000
Total Assets		$1,100,000

Liabilities

Current Liabilities		
Accounts Payable		$200,000
Taxes Payable	44,000	
Debts Payable	6,000	
Expenses Payable	40,000	
Total Current Liabilities		$290,000
Long Term Liabilities		
Mortgages	$ 90,000	
Notes	40,000	
Total Long Term Liabilities		$130,000
Total Liabilities		$420,000

Net Worth

Capital Stock	$200,000	
Retained Earnings	280,000	
Current Earnings	200,000	
Total Net Worth		$680,000
Total Liability and Net worth	$1,100,000	

Typical Balance Sheet

Chapter 21

Contracting for Profit

WHAT DETERMINES YOUR PROFITABILITY AN OVERALL VIEW

Productivity

The number one major factor that determines your profitability is your productivity rate. For sheet metal contractors the rate of fabrication and installation, what machinery, tools and equipment are owned, the methods used to operate and your quality of manpower are primary factors. Do you fabricate at 22 pounds per hour or 28, or at the average 25? Do you have a duct coil line or a hand break? Productivity between companies can vary 30 or 40 percent. For piping contractors, the rate of erection, field efficiency and quality of the manpower are primary productivity factors.

Being On Top of the Job

Avoiding costly delays, goof-ups, miscoordination, out of sequence work, over manning, waste; the control of all the operations, of timing, being on top of costs, confronting problems as they come up, is the 2nd major factor that controls your profitability. The ductwork or piping for the 2nd floor is two weeks late and it costs an extra $1,000 to install because the other trades got in first. Or some equipment doesn't fit and it costs you $2,000 to rectify the problem.

Complete and Accurate Estimating

You can be off 10 to 20 percent without batting an eye lid. The accuracy of your estimate is a major, constant factor in determining the profitability of your company. Treat each bid like a new born baby, give it the time and attention it must have to survive.

Correct Markups

Hand in hand with estimating direct costs is the proper evaluation of overhead costs, setting of profit goals, and the adequate application to each bid. The markup money which isn't covered for overhead comes off the top, which is the profit layer.

Buyout and Good Purchasing

The fifth most important factor which determines your profitability is buyout and frugal purchasing. If you bid a job with no commitments to equipment suppliers or to subcontractors, you just accept their bids as received, make no deals, you can well buyout with a 3, 5 or even 10 percent pickup, with a more thorough evaluation after you receive the contract, than is possible at bid time.

Many equipment and sub quotes are too high going in, not really competitive and some have contingencies. If you allow your suppliers and subs to reevaluate their pricing, recheck their take-offs and extensions before purchasing from them, you frequently uncover errors and inflated pricing. You may develop cost saving ideas with the additional time and the second look, which can reduce costs. Two suppliers may have bid $22,000 and $24,000 and later found they could trim and polish their bids for something under $20,000. You get several prices on 50,000 pounds of galvanized ductwork and you find some one who can sell it for less per pound.

Selective Bidding

Another factor that has a major influence on how well you operate, and consequently how you fare economically, is selective bidding: choosing those jobs you can do well; volume you can handle; projects where you have a history of experience and pricing on; jobs that suit your operations in terms of facilities, machinery, manpower, and management.

Do you do well with smaller, shorter simpler projects with multi-skilled tradesmen, and with service operations, or is your expertise with large and medium size projects that span 1, 2 or 3 years and which involve highly sophisticated production equipment in your shop, aggressive foremen on the job site, and a wide range of special skills.

Selective bidding involves being correctly capitalized for a project and not having to go out and borrow $100,000 or $200,000 at 12% to run a job. The amount of money you must or must not borrow effects your profitability potential.

Timely Payments

Timely payments for work you have completed on projects is another factor in your profit making picture. If you have to wait 6 months, or a year for $20,000 or $50,000 and it's worth 8 or 10 percent per year, you are hurting money wise and cash flow wise. Waiting 60 or 90 days or more for monthly progress payments are damaging to your financial situation. Cash flow control is an absolute determinant on your profitability, and prompt invoicing, follow up and demand of payment within a reasonable time span is a necessity.

Controlling Overhead

Overhead can effect you astronomically if left to its own meandering. A $500,000 overhead can jump to $600,000 or $700,000 in a year or two without any increase in business or work load.

Job Costing

Job costing and time studies provide the factual figures for estimates, for production scheduling, progress billing and valid financial reports and can indirectly effect efficiency and accuracy.

Inventory Control

Not having $40,000 worth of uninstalled fans on the job site 6 months, where you have to pay the manufacturer, but you can't collect from the owner; or an over abundant supply of galvanized material in the shop, can chew away the dollars that should be profit in the end.

Bidding and Pricing Strategically

Bidding strategy can often provide many extra thousands of dollars by not leaving too much on the table, by properly sensing the market and the competition.

Being On Top of Finances

Job costing reports and control, regular financial statements and managing cash flow effectively are all essential for bottom line profits.

HOW TO LEGITIMATELY
REDUCE COSTS ON A BID

To Be More Competitive

System and Design
1. **Redesign**

An intelligent redesign or change in specifications while still giving the customer the same performance at a substantially lower price can open many

doors for you. Redesign should always be a consideration when looking for ways to reduce costs.

2. **Lower Priced Systems**
   - A roof top or split DX system might handle the customers needs just as well as a chilled water system and costs considerably less.
   - Combining a number of small systems together into a larger one can often reduce costs.
   - Using a 2 pipe system versus a 3 or 4 pipe can save if feasible.
   - Eliminate or minimize return air ductwork if possible.

3. **Check for Over-Design**
   A system may be over designed, too full of design safety fudge factors, excessive capacity, larger sizes than need be, and can be reduced to just what is really needed providing substantial cost reductions.
   - Where 40 tons of cooling will do the job on a project instead of an excessive 50 tons, costs can possibly be reduced 10 or 20 thousand dollars.

Ductwork and Piping
4. Reduce Ductwork and Piping Costs Through "STAR" Method of Streamlining
   - Straighten out ductwork and piping routings.
   - Maximize duplications of runs and fittings.
   - Optimize arrangement of straight and fittings in runs.
   - Simplify and standardize fittings.

   Five hundred feet of ductwork might be cut down to four hundred with more direct and efficient routing, outlet placement and so on. This can well reduce costs on a $100,000 project to $86,000, a sure winner to get you a job once substantiated.

5. **Lower Priced Ductwork or Piping Materials**
   Different materials which work equally as well as those specified, but cost less, might be offered as a substitution on a bid.
   - 24 gauge cleated stainless steel ductwork may be equally resistant to the fumes in a system as PVC ductwork, and costs 20% less.
   - Threaded black steel pipe and fittings may be 25 to 30 percent less than copper to purchase, and still work equally well in certain applications.

Equipment
6. **Lower End Standard Equipment and Systems Versus Deluxe**

Deluxe equipment, materials, control systems, construction, accessories, options, etc. may be as extravagant as a Mercedes, whereas standard features and design may be more than satisfactory at a much lower cost.

- Use XYZ equipment at $9,000 instead of ABC equipment at $12,000.
- Use an electric control system versus a deluxe pneumatic.

7. **Substitute Lower Price Non-Specified Equipment**
Substitute non-specified equipment manufacturers with a lower priced manufacturer who can meet the performance and construction of the specified equipment. This can make the difference in the final low bid price.

8. **Pre-Negotiated Equipment and Sub-Contractor Prices**
Equipment and sub contractor prices can be negotiated at the time of bidding. Agreements can be made and prices might be considerably lower than street prices. This frequently provides a competitive edge.

9. **Package Buy at a Discount**
You may have several projects that will be going at the same time and can package buy equipment and materials at a volume discount, thereby edging under competitors who are pricing the same items in the bid at the street quotations.

Grilles and diffusers for three projects all bought simultaneously from the same supplier might be five or ten percent cheaper.

Labor
10. **Use Lower Priced Labor**
Where *conditions permit* you might use helpers, stockmen, truck drivers, draftsmen, etc. at lower wage rates which may tilt the scales in your favor when bidding.

Machinery and Methods of Operation
11. **More Efficient Productive Machinery or Methods**
You may plan to buy a high speed forming machine, coil line, plasma cutter, heavy duty power shear or press break or some other equipment which will increase efficiency. This can be factored into your bid to lower the price and help you be more competitive.

Estimates
12. **Recheck Estimate for Overage Errors**
Errors in estimates are made in both directions. You may have $5,000 too much in your bid as well as too little due to math or takeoff errors or poor pricing. This throws you out of the running and you don't get the job. Recheck estimates thoroughly, not only for loss errors but for overage also.

STAR METHOD OF REDUCING
DUCTWORK AND PIPING COSTS

**Standardize and Simplify Ductwork and
    Piping Designs**
Create Duplications
Arrange Pieces More Economically
Route Runs More Efficiently
Optimize Ductwork and Pipe Sizing

You design and lay out an air distribution system and find that there are 300 pieces of ductwork in it. The current market price for the average size duct section in this design, let's say, is approximately $150 each. This brings the total price for 300 pieces to $45,000.

Now you take the initial layout and apply the principle rules of STAR to it for reducing costs. You then recount the pieces and discover that you've reduced the number from 300 to 250. You also re-evaluated the cost of each piece and determined that the average price is reduced 10 percent also, from $150 to $135. Consequently, you have slashed the total installed cost of the ductwork 25 percent, a $15,000 saving, from $45,000 to $30,000.

Can this be true, you wonder? So you check whether the calculated performance has been affected and find it unchanged. If anything, the total static pressure drop is lower, the noise level unaltered, air distribution in the spaces still as intended, and anticipated air volume delivery still on target.

How can you do this with the same ductwork materials, the same designer, the same contractors and sheet metal men, and the same architectural design?

STAR is the answer, a value analysis approach for reducing total costs while still maintaining function and performance in building piping and ductwork systems. The mnemonic acronym is derived from the four basic elements of the method: S for standardization, simplification and sizing, T for typicals, A for arrangement, and R for routing. But more on these later.

STAR is a method of vigilant search for, and extermination of, costs that do not contribute to a system's objectives and to the user's needs. Its principles are aimed at maximum compatibility between performance and cost efficiency in any pipe or duct conveying system, whether air, liquid, steam, electricity, or otherwise. Its specific goals are fewer joints of pipe and fittings, more standardized economical ones, more direct routing, and fewer connections, so that labor time is reduced. Material is generally considered secondary in this approach.

STAR is a process of revising and polishing, of editing, of whittling away at what is unnecessary and creating more economical alternatives.

STAR helps contractors to meet their estimates and to avoid losing money in a hotly competitive market with pricing so low that firms cannot afford to manage as they should. The design engineer benefits from STAR in that he can get more work through lower cost systems, and he derives the professional satisfaction of designing more efficient systems.

The owner has the most to gain from STAR, however, he gets more for his construction dollar; it's easier for him to meet budget allocations; there is generally a reduction in operating costs; his building gets constructed quicker because of the reduced labor hours; and his initial outlay is less. And if he chooses, he can get additional features in his systems with the money saved through STAR. In some cases, it might enable those who could not otherwise afford to build a chance to do so.

Why Extra Costs Are Built In

Why aren't all duct and pipe systems designed, fabricated, and installed in the most economical and efficient way? There are at least three basic reasons:

1.   The first is a lack of detailed cost knowledge by the designers and even by contractors themselves.
2.   The second is that designs are assumed to be the most economical when they're taken off the boards and sent out for bidding.
3.   And third, repeated and objective checking of completed designs to ferret out and eliminate unnecessary costs that have been built in isn't done often enough if at all.

Design is a complicated, creative process, involving many engineering considerations and calculations. Detailed cost reduction can be done more effectively after the design is completed by looking at the entire layout. The consideration and creation of alternate options become more concrete when you have a layout in front of you.

There are additional reasons why duct and pipe systems aren't put in at optimum cost: projects are rushed; experienced and qualified personnel aren't available; and the additional man-hours of engineering required to trim

costs are not always in the interest of those working on a percent of construction fee basis.

Another major reason is that designers do not work closely enough with contractors in the design stage. Contractors can provide needed information on detailed costs, results of installation, performance of systems, and problems encountered and solved along the way.

The last category of causes for excessive costs includes "shooting from the hip," "budget load calculation" designs, fudge factors, and contingencies in lieu of precise, thorough, and complete design calculations.

What is Value Analysis?

What is value analysis really all about? Here's an example of one aspect of it. Max Smart, an accountant, wants to buy a car. He's married, has one child, and doesn't plan to have any more. His objective is to get reasonably comfortable, dependable, quality transportation for three persons for short distance driving.

Max looks at a $40,000 Cadillac first, and this is what he'd love to have most. But he sees there are many unnecessary costs built into this luxurious, powerful, and spacious automobile that he just doesn't need. So next he looks at an Oldsmobile. It's sure nice, too, Max thinks, but it's the same situation: he just doesn't need high horsepower, power windows, etc. How about a full size Ford or Chevrolet? Naw, he'd still be paying for more pounds of steel and more cubic feet of space than he needs.

Compacts—they're better, and why not go all the way and get a car that efficiently uses the energy put into it, has no wasted internal space, is highly reliable, and will have no unused horsepower. A Toyota sub-compact fits the criteria, he decides, and that's what this objective, analytical gentleman buys, for $12,000. This is $28,000 less than the Cadillac, and Max gets all the satisfactory performance he needs without spending money unnecessarily.

The point here is not to recommend the purchase of an economical foreign car. Rather, it's that unnecessary luxuries, over seeing, inefficient usage, conspicuous waste, and complexity of products that frequently confront us in our lives can double, triple, and quadruple the costs of things without giving us the returns we need or want for the extra costs.

Extra Costs: Where to Look

There are many categories of unnecessary costs, in addition to those just mentioned, that STAR methodically attacks. A few of the most important ones are:

Using many parts where a few or only one will do the job.

Using many operations, activities, or steps; often an extensive series can be drastically reduced.

Choosing the first solution or design without creating or considering alternatives.

Not having all the pertinent information or necessary facts.

False cost savings, where shaving something in one area causes total costs to increase, such as attaining lower initial cost through a measure that will increase operating cost inordinately, or vice versa.

Attitudes Don't Help

We also have attitudes that often hinder our application of value analysis and consequently prevent us from reaching an ultimate cost reduction. Have you ever heard any of these remarks?

"If it doesn't work, what will happen to me?"

"This isn't the way we do it."

"I know I'm right."

"It won't work!"

"I'm the boss-we do it my way."

"This is what the plans and specs call for, and that's the way you're going to do it."

"What will people think about this?"

"This is the way I was told to do it."

"You can't do it that way. All you want to do is save yourself money; you don't care what the customer ends up with."

"I don't feel like changing."

"There's no benefit in it for me."

"It's to my disadvantage."

Beware of these attitudes in yourself and others, and don't let them sway your objectivity and good judgment.

Knowledge Required

STAR requires that you possess, whether you are a designer or contractor, the following:

- A knowledge and understanding of heating, ventilating, and air conditioning systems.
- A familiarity of the results that must be achieved from these systems.
- An awareness of the "must" objectives of the systems.
- A knowledge of the various means that will achieve the desired results.
- A knowledge of the factors that can affect the results.

- An awareness of the criteria ranges or tolerances of these factors.
- Information on total costs, unit budget costs, and detailed costs.

Overall HVAC General Results and Criteria Required

The overall general results that must be achieved through heating, ventilating, and air conditioning systems are:

- Physical comfort.
- Healthy air conditions.
- Air conditions in which persons can perform work properly and efficiently or do whatever it is they are to do in the space.

The criteria that must be met in achieving these goals include:

- Reasonable initial and operating costs within the budget.
- A system that will work with variable weather conditions.
- A system that will work with variable space usage conditions.
- Reasonable maintenance requirements and adequate accessibility.
- Reasonable dependability and reliability.
- A system that fits reasonably well into the architectural design.
- A system that meets applicable building codes.

Specific Objectives

Tolerance ranges of various parameters that we will operate within and use as guides include:

- Air movement, local air velocities of 10 to 45 fpm.
- Dry bulb temperature, 73 to 75°F.
- Relative humidity, 20 to 60 percent.
- Mean radiant temperature, 70 to 80°F.
- Sound level, 30 to 35 dB.
- System static pressure, within 15 percent of design.
- Air volumes, within 5 percent of design.
- Duct velocities, 400 to 1600 fpm for low velocity systems in commercial and institutional buildings.
- Duct static pressure drop, 0.01 to 0.3 in. WG per 100 ft for low velocity systems.
- Temperature control, space temperature within ± 1°F of design.
- Temperature drop or gain in duct runs, ± 5°F of design temperature at outlets for single zone systems with controls at supply unit (this is no problem with terminal reheat, dual duct, and hot and cold deck multizone systems because the final discharge temperatures are modulated and controlled).

• System balance, within 5 percent of design.

Acceptable Duct Static Pressure Increases

The static pressure drop of a trunk or branch duct run can be increased, if need be for initial cost economy, to match that of the run with the highest static pressure drop in the system (provided, of course, that the run so handled is not part of the run with the highest static pressure drop). It is the same as a critical path schedule. You can increase the time for any activity that is not on the critical path, up to the point where the time for its path equals the critical path's.

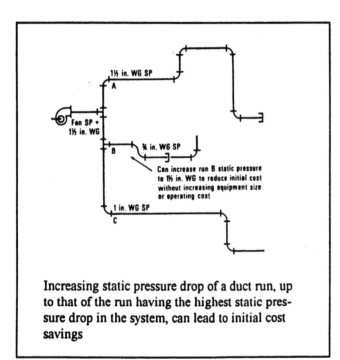

Increasing static pressure drop of a duct run, up to that of the run having the highest static pressure drop in the system, can lead to initial cost savings

This can vary with whatever safety factor the designer adds to his static pressure calculation, with difference between brake horsepower and actual horsepower of the motor selected, and with the actual performance of the equipment. Keep in mind that an increase of 15 or 2 0 percent in system static pressure drop—with no change in fan size, drive speed, or horsepower—may not change the air volume more than 6 or 7 percent. And a 1°F increase in air temperature will overcome this airflow decrease.

Critical Times to STAR, Definitions

As you design and lay out ductwork and piping systems, you will, of course, use all your knowledge of cost saving techniques without letting this distract you from effective engineering. There are four critical times, however, when you should concentrate your attention and efforts to the costs of the system without design thoughts prevailing.

These are:
• After your rough schematic design is completed.
• After the 1/8 inch scale design drawing is laid out.
• After the contractor's 1/4 inch scale background drawing is completed.

As indicated previously, the S in STAR stands for standardization, simplification and sizing. Look for opportunities to standardize the design, sizes, lengths, and component parts of the ductwork. Simplify areas, runs and designs.

The T stands for typicals. Compare runs, pipe sections, and fittings for duplications, either in their entirety or in their component parts. Compare parts of a run itself for typicals, runs with other runs, areas of a building with other areas, and one project with other projects.

The A stands for arrangement. Look for improvements in the arrangement of pieces in runs, which might include different lengths, different combination of pieces, and elimination of fittings.

The R stands for routing. Check each trunk run, branch, and riser for routing improvements that may lead to more direct routing and more efficient equipment locations.

These are the STAR procedures. Following is a more detailed guide to their implementation.

DETAILED APPLICATION

Standardization, Simplification and Sizing

The more variations one has in dimensions, design, configuration, and performance, the more labor time, complications, errors, lack of control, and confusion one will encounter, all of which lead to higher total cost. Standardization leads to simplification, efficiency, cost reduction, interchangeability of parts, and mass production.

Standards are not haphazard or arbitrary. They are established to make sure things fit together, to assure they perform as expected, to avoid confusion, and to prevent recurrence of past errors.

The following methods can be used to standardize and simplify duct runs and consequently reduce costs without impairing system performance:

1. Make taps in tees one length and one configuration.
2. Standardize elbow radii to size ranges.
3. Use a single section offset rather than two elbows and pipe.
4. Standardize the length of straight duct sections.
5. Make change fittings constant in depth so they can

be fabricated from two pieces instead of four.

6. Reduce the number of different duct sizes; stay away from odd and excessively large sizes.
7. Use standard size panels in built-up housings.
8. Avoid seaming on flat sides of ducts wherever possible.
9. Avoid double curved offsets.
10. Avoid changes in depth at elbows.
11. Avoid changes in depth at divided flow wye fittings.
12. Design for mass production runs.
13. Use duct sizes that come from standard sheet sizes.
14. Avoid cramped, messy criss-crossings.
15. Minimize the use of fittings; shoot for as much straight duct as possible.

Convert Rectangular Ductwork to Round Ductwork

Round ductwork is the most efficient ductwork in regard to performance such as maximum air flows per surface square feet and per the least amount of resistance per running foot. Flat rectangular ductwork with width to depth aspect rations of 2, 3, or 4, restrict air flow and can develop pressure drops that are twice that of round duct. Hence, converting rectangular ductwork to equal friction round, can save 20 to 30 percent in square foot of ductwork materials required per the same amount of air delivery.

This application works best with duct branches and small main runs. Larger rectangular main duct runs generally become too large in round sizes to fit into ceiling spaces, etc.

The method of determining the equal fiction round sizes in lieu of the given rectangular size is based on:

- Determining the actual friction factor per 100 feet for the rectangular size per ASHRAE manuals first.
- Then working backwards from the friction factor to the equivalent round diameter. The following are some examples.

Rectangular Size	Equal Friction Round Diameter
12 x 6	9
18 x 6	11
24 x 6	12
24 x 12	18
36 x 12	22
48 x 12	25

WinDuct Estimating System Does Automatically

However, there is no need to do this by hand. The WinDuct Wendes estimating systems performs the rectangular to round calculations and selection of optimal rectangular sizes automatically.

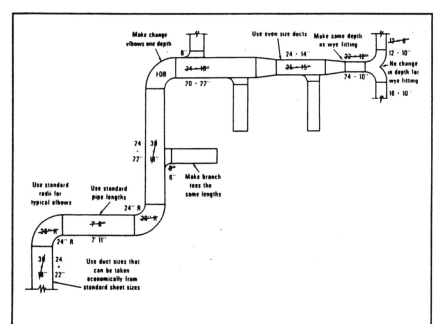

Standardization and simplification are important procedures in STAR value analysis. Study of the changes in this air distribution system reveals specific methods of standardization, which leads to simplification.

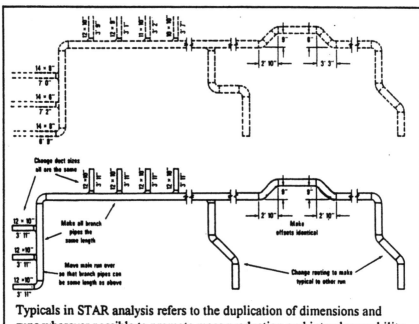

Typicals in STAR analysis refers to the duplication of dimensions and runs wherever possible to promote mass production and interchangeability of parts. Comparison of original and revised systems here shows specific methods.

Typicals or Duplicate Fittings

This involves duplications, interchangeable component parts, and mass production runs for both fabrication and installation. Here are some methods of eliminating unnecessary costs in the typicals department:

1. Make a series of branch ducts typical in size and length.
2. Reroute a duct or pipe run to make it typical with one or more other runs.
3. Make duct section lengths the same.
4. If you have two offsets or two change fittings with the same dimensional changes, make them the same length.
5. Move a trunk run to make the branch-offs typical with a series of branch-offs on another run.
6. Change duct sizes, within the limits of design criteria tolerances, to make them typical with sizes in other runs.
7. Make size changes also to create interchangeable parts.
8. Make divided flow wye fittings typical.
9. Elbows and tees offer marvelous opportunities for duplication.
10. Search for and create typical pieces, runs and areas.

Arrangement

You can juggle pieces in a duct or pipe run, alter lengths, combine pieces, eliminate them and build subassemblies without adversely affecting system performance. In fact, it is more likely that performance will benefit from these measures. Here are some specific methods:

1. Combine an offset and a change fitting into one piece.
2. Combine short pieces of duct.
3. Combine a size change fitting with an elbow if there is no depth change.
4. Combine two 90' Elbows into one offset.
5. Combine two elbows and the piece of pipe between into one offset.
6. Eliminate change fittings and maintain the larger duct size as long as possible.
7. Alter lengths of pieces and juggle their locations back and forth until you have the fewest pieces and connections possible in a run.
8. Assemble duct sections in the shop whenever practical.
9. Substitute a shop welded connection for a field companion angle one.
10. Make as many connections as possible on the floor at the job site before hanging.

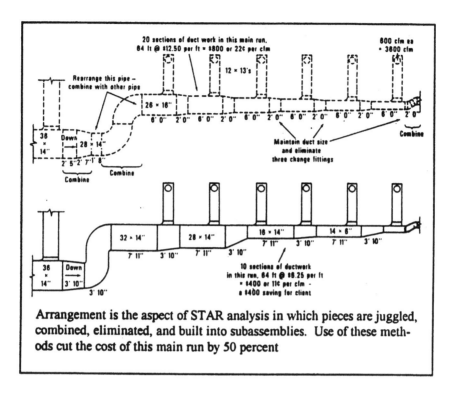

Arrangement is the aspect of STAR analysis in which pieces are juggled, combined, eliminated, and built into subassemblies. Use of these methods cut the cost of this main run by 50 percent

Routing

Direct routing and strategic placement and orientation of equipment can be one of the major areas of cost reduction, not only for air distribution systems but piping and wiring systems as well.

The ultimate in efficiency of routing, air distribution, and equipment location is found in large single story discount stores with exposed duct runs. The air handling units are at the back of the store, their discharges facing the front. Ductwork is routed straight out of the units, directly across the store, with straight duct sections and only one or two change fittings. There are no turns or offsets. There is a grille at the end of the run to blow air on the front windows and a few grilles flush on both sides of the ductwork between there and the unit.

Skeptical though some may be, if the engineering is done correctly, the system achieves and maintains the objectives we are committed to. Physical comfort is provided for customers and employees, air conditions are healthy and people can function as required.

The cost of the ductwork in this case is approximately 10 percent of the ventilation cost pie, not 30 or 40 percent as is the normal range.

Here are specific methods and things to watch for in routing distribution and locating equipment:

1. Avoid loop routing wherever possible.
2. Avoid step routing.
3. Route directly at an angle instead of making turns.
4. Combine parallel runs if possible.
5. Move a main run over ceiling outlets and eliminate branches.
6. Move a main run closer to outlets to shorten branches.
7. Route branches directly to outlets or terminals.
8. Route runs directly without offsets wherever possible.
9. Raise or lower criss-crossing ducts or reduce duct depth to eliminate multiple offsets and elbows.
10. Avoid cramped or unnecessary criss-crossing by rerouting.
11. Keep runs at the same elevation as much as possible.
12. Place mixing boxes between hot and cold trunk runs to avoid criss-crossing branches and trunk runs.
13. Turn equipment so that duct or pipe connections are perpendicular to the runs they attach to, thereby eliminating turns. Provide straight shot orientation.
14. Locate equipment at the same elevation as the duct or pipe runs, if possible, to avoid risers, offsets, turns, etc.

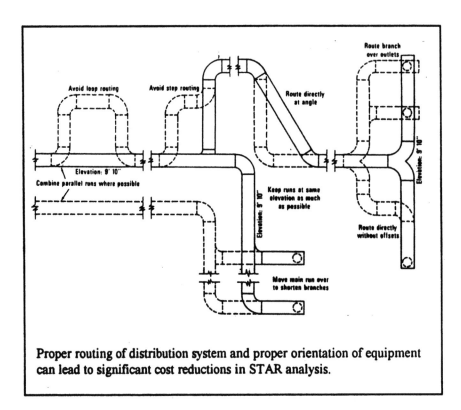

Proper routing of distribution system and proper orientation of equipment can lead to significant cost reductions in STAR analysis.

Implement STAR as a Standard Procedure

STAR can be implemented over and over. You will continually find more and more unnecessary costs built into mechanical systems. The reasons for them are immaterial; they're there. And you'll find yourself and others through your effort-through using your imagination and creative abilities and through the valuable technique of simply looking.

ASHRAE Incorporates STAR Techniques

Since these STAR techniques and principles were developed by Herb Wendes, and this article was originally published in HPAC magazine in 1979, ASHRAE has incorporated many of the methods of cost reduction into their manuals.

Chapter 22

Computerized Estimating Principles

Wendes Systems, Inc. is pleased to announce two new powerful Windows based estimating systems that incorporate the principles and methodology described in this manual in an easy-to-use computerized software program. The systems are easy to learn and are ready to use with preloaded materials and labor.

The WenDuct sheet metal estimating system is fast, accurate and easy-to-use with pre-loaded price and labor tables. A customizable database allows for maximum flexibility. Computerized Takeoff incorporates fast digitizer technology, electronic pen, material templates and mouse driven point and click menus. The accurate Wendes method of estimating produces extensive reports and a bottom-line bid directly from the takeoff. The system is flexible with user-definable square footage equations, customizable reports, tables and specifications for material and labor. Choose between a variety of duct types within a takeoff by selecting galvanized, spiral, industrial and fiberglass systems.

After a takeoff is completed a system may be "value engineered" by comparing alternate duct types and materials. The software will automatically re-calculate an entire system based on alternate duct types and materials calculating new material and labor. The calculations for ductwork and fitting are converted based on the aspect ratio of the original takeoff.

WenDuct will automatically produce a detailed estimate by accepting the transfer of data from a wide range of Computer Aided Drafting and Detailing Systems. An estimate may be easily produced based on the as drawn system vs. an original bid. Changes in job scope may be more accurately tracked and change orders produced. Data is available for viewing real time during takeoff, per selected item or system.

The new WenPipe piping and plumbing estimating system includes an extensive 60,000 item database covering a wide range of HVAC, plumbing, piping, process and industrial materials. Computerized takeoff incorporates fast digitizer technology, electronic pen, material templates and mouse driven point and click menus. Accurate Labor is based on national labor or MCA labor available to registered MCA members. Extensive labor and material reports are produced directly from the takeoff. Harrison Pricing Services and Wendes Systems provide a seamless interface that links the estimating material database with updated Harrison pricing. Ferguson Enterprises, Inc. and Wendes Systems, Inc. have partnered to make Ferguson Commodity Pricing available to Ferguson customers. Ferguson Commodity Pricing is available directly through Ferguson Enterprises. Other pricing services are available for all North American Regions.

For detailed information or to view the capability of computerized estimating contact the Wendes Systems, Inc. technical and training staff at 1-847-808-8371. Software is available on a free 30-day evaluation basis. On-line training is provided with evaluation software. Additional information is available on the Wendes Systems, Inc. web site: *www.wendes.com.*

For Product Safety Concerns and Information please contact our EU
representative GPSR@taylorandfrancis.com
Taylor & Francis Verlag GmbH, Kaufingerstraße 24, 80331 München, Germany

www.ingramcontent.com/pod-product-compliance
Ingram Content Group UK Ltd.
Pitfield, Milton Keynes, MK11 3LW, UK
UKHW052038180425
457613UK00024B/1284